National Academy Press

The National Academy Press was created by the National Academy of
Sciences to publish the reports issued by the Academy and by the
National Academy of Engineering, the Institute of Medicine, and the
National Research Council, all operating under the charter granted to
the National Academy of Sciences by the Congress of the United States.

ESTIMATING POPULATION AND INCOME OF SMALL AREAS

PANEL ON SMALL-AREA ESTIMATES OF
POPULATION AND INCOME
Committee on National Statistics
Assembly of Behavioral and Social Sciences
National Research Council

NATIONAL ACADEMY PRESS
Washington, D.C.   1980

Library of Congress Cataloging in Publication Data

National Research Council.   Panel on Small-Area
   Estimates of Population and Income.
   Estimating population and income of small areas.

   Includes bibliographies.
   1. Population forecasting—United States.
   2. Income forecasting—United States.   I. Title.
   II. Title: Small areas.
HB849.53.N37   1980   001.4′22   80-26012
ISBN 0-309-03096-X

Available from

NATIONAL ACADEMY PRESS
2101 Constitution Avenue, N.W.
Washington, D.C. 20418

Printed in the United States of America

PANEL ON SMALL-AREA ESTIMATES OF POPULATION AND INCOME

iii

iv

Contents

List of Tables

Preface

The Bureau of the Census of the U.S. Department of Commerce has long had an interest in developing postcensal estimates of population for areas smaller than states. Until the 1970s, its population estimation program was limited to counties, large cities, and metropolitan areas. During the 1970s, however, the Census Bureau undertook the major task of making estimates of population and per capita income for some 39,000 general purpose local jurisdictions. This undertaking was stimulated by the State and Local Fiscal Assistance Act of 1972 (P.L. 92-512), commonly referred to as general revenue sharing (GRS), which requires that the most recently available data provided by the Census Bureau be used to determine the allocation of GRS funds among the states and approximately 39,000 eligible units of local government—counties and subcounty areas.

The Census Bureau's program of estimates are important not only because large amounts of federal funds are allocated directly on the basis of those estimates but also because population estimates are basic to other measures, such as current vital rates. Planners and decision makers at the state and local levels also rely heavily on the small-area estimates.

At the request of the Census Bureau and the Office of Revenue Sharing of the U.S. Department of the Treasury, the Committee on National Statistics in July 1978 established the Panel on Small-Area Estimates of Population and Income. The Panel included persons with expertise in the areas of statistics, demography, and economics. (Biographical sketches of Panel members appear in Appendix L.)

The Panel was charged with the general task of evaluating the Census

Bureau's procedures for making postcensal estimates of population and per capita income for local areas. More specifically, the Panel was asked to review methods currently used and possible alternate methods, review data sources currently used and possible alternate sources, and assess levels of accuracy of current estimates in light of the uses made of them and of the effects of potential errors on these uses.

In carrying out its task, the Panel was asked to develop its recommendations in the light of the 5-year schedule for future censuses and available information on the census undercount; consider criteria for choosing among data sources and techniques—for example, the importance of uniformity and consistency in order to treat different localities equitably—and the standards of accuracy required for places of different sizes; consider the error structure inherent in the estimates, how estimates of error might be prepared, and how (if at all) such estimates might be conveyed to users; and consider the appropriate role for State agencies in cooperating in the estimating process.

Because a complete description of the detailed procedures used by the Bureau to prepare the estimates of population and income was not available in written form, the first task undertaken by the project staff was the preparation of Appendix A, "Postcensal Population Estimation Methods of the Census Bureau." Although the authors spent a considerable part of 2 months at the Census Bureau preparing this appendix, it is not an official report of the Census Bureau, and some of the minute details of the procedures (which are based on written census reports, supplemented by discussions with census staff) may not, despite the authors' efforts, be described exactly as actually carried out during the 1970s. A similar qualification applies to Appendix B (the summary of income estimation methodology) and to the statements in Chapter 1 concerning the rationale of the methodology, the criteria of accuracy used, and the reasons for the methodological decisions made by the Census Bureau.

The Panel acknowledges with gratitude the assistance received from many individuals who cooperated in the study: Meyer Zitter, Roger Herriot, Richard Engels, Mary Kay Healy, and Robert Fay of the Census Bureau; Matthew Butler, Kent Peterson, and Jack McGuire of the Office of Revenue Sharing; Joseph Duncan of the Office of Federal Statistical Policy and Standards and Edwin Colemen of the Bureau of Economic Analysis of the U.S. Department of Commerce consulted with members and staff on several occasions. Many other members of the Census Bureau provided assistance, and special thanks are due David Word, Frederick Cavanaugh, David Galdi, Joseph Knott, Sharon Baucom, Jerome Glynn, Richard Irwin, Jennifer Marks, Edward Hanlon, Marianne Roberts, Barbara van der Vate, Louisa Miller, Frances Barnett, Joel Miller, and Marshall Moore. Martin Ziegler and Frederick Cronkhite of the Bureau of Labor Statistics, U.S. Department of Labor, and George Sturm of the Bureau of Health Planning, U.S. Department of Health, Education, and Welfare, were helpful in explaining uses of the postcensal estimates by their respective agencies.

Members and staff of the Committee on National Statistics provided advice at many phases of the Panel's work. Margaret Martin, past executive director, Edwin Goldfield, executive director, and Miron Straf, research director, were generous with support, criticism, and guidance.

Above all, the Panel wishes to acknowledge the major contribution of the project staff. Bruce Spencer, study director, had overall responsibility for coordinating the work of the Panel and the staff, and he made important contributions to every phase of the study. He provided the working materials for the Panel, organized its meetings, prepared many of the background papers that served as the basis for our discussions, and was largely responsible for drafting this report. Che-Fu Lee and the late Walt R. Simmons also contributed to parts of the project. Linda Jones was secretary for nearly all of the Panel's duration. We also acknowledge the superb editing skills of Jean Savage, Elaine McGarraugh, and, especially, Genie Grohman.

Finally, I wish to thank the members of the Panel for their willingness to contribute their time and specialized knowledge to the tasks assigned us. A number of Panel members prepared background papers for our discussions. Some of their contributions appear in the appendices; others have been incorporated in the text of the report. This report represents the consensus of the Panel on the issues addressed. Needless to say, however, no individual member of the Panel should or would want to be held responsible for every detail or point of view expressed.

EVELYN M. KITAGAWA, *Chairman*
Panel on Small-Area Estimates of Population and Income

PART
I
SUMMARY
REPORT

1

Overview
and
Recommendations

1.1 INTRODUCTION

1.1a BACKGROUND

The decennial census provides counts of the enumerated population[1] and estimates of per capita income for detailed geographic areas of the United States at 10-year intervals. In the years following a census this information becomes outdated as the population and per capita income of areas change. The objective of the Census Bureau's postcensal estimation program is to update the census information on population and per capita income for approximately 39,000 general purpose governmental units, more than half of which have populations of less than 1,000.

The preparation of postcensal estimates for those many small areas was prompted by the State and Local Fiscal Assistance Act of 1972 (P.L. 92-512), which required that the most recently available data on population and per capita income provided by the Census Bureau be used in the formulas that determine the annual (or biannual) allocation of general revenue sharing (GRS) funds among 39,000 eligible units of government. In addition to their use in determining the allocation of GRS and other

---

[1] Although the population counts derived from the decennial census enumerations are designed to be complete and accurate, they are known to contain errors of omission, duplication, and misclassification. The Census Bureau publishes estimates of the net undercount of the population by sex, race, and age for the United States as a whole (Bureau of the Census, 1973a).

3

federal assistance funds—a total of more than $36 billion per year—these estimates also serve a wide variety of needs of state and local governments, private organizations, and scholarly research (see section 1.1b).

Although the Census Bureau had been working on methods for estimating the population of states and large counties and cities since the 1940s and had published its first series of estimates for all counties in the United States in 1966, the methodology for small areas was in the early developmental stages when the 1972 act was drafted. In a hearing before the Ways and Means Committee, Census Bureau officials stated that the methodology for producing estimates for small local areas was not yet developed and tested and could be very inaccurate for places of population under 50,000 (U.S. Congress, 1972). In drafting the legislation, Congress did not require that postcensal estimates be produced regardless of accuracy but only that the most recent data provided by the Census Bureau be used for general revenue sharing allocations. Especially for population estimates at the subcounty level and for estimates of per capita income, the methodology currently used by the Census Bureau to make postcensal estimates was developed to a great extent after the GRS law was enacted.

The Panel's review of the postcensal estimation program of the Census Bureau has included an examination of the logic and the accuracy of the methods used to derive estimates of population and per capita income. Our tentative assessment of their accuracy is based primarily on comparisons of the postcensal estimates with the results of special censuses carried out during the 1970s. More conclusive evaluation awaits comparison of the estimates with the results of the 1980 decennial census.

Although the postcensal estimation program produces estimates of total population and per capita income for approximately 39,000 areas, the estimation methodology is designed to measure the *change* in total population and per capita income of each area since the last national census enumeration. The estimates of change for each area are applied to its population and per capita income as determined in the last census. Thus the implicit objective of the methodology is the estimation of postcensal change in population and per capita income, and the Panel has evaluated the methodology from this perspective as well as in terms of the accuracy of the estimates of total population and per capita income. The accuracy of estimates of postcensal change is of critical importance when the postcensal estimates are used to calculate the allocation of general revenue sharing funds, because it is changes (since the last census) in population and per capita income of areas that produce changes in the allocation of funds.

The Panel has not addressed the question of whether or not the postcensal estimates of population should be adjusted for census undercount for two reasons. First, this question is equivalent to asking whether or not reported census figures should be adjusted for census undercount (since census data and postcensal estimates must be consistent in this respect), and this was recently considered by another panel of the National Research Council (1978). Second, if a decision were made to adjust the population estimates for census undercount, essentially the same postcensal methodology currently used by the Census Bureau could be used to estimate postcensal change; the major difference in procedure would be that the reported data from the last census would be adjusted for census undercount before being added to the estimates of postcensal change.

1.1b NEEDS FOR POSTCENSAL ESTIMATES

The Census Bureau currently produces postcensal estimates of population and income for approximately 39,000 general purpose governmental units that are eligible for general revenue sharing funds. Table 1.1 shows the numbers of county and municipal and township governments and their estimated population, classified by size. The overwhelming majority of these areas have very small populations. For example, 85 percent of the 35,684 municipalities and townships had less than 5,000 population in 1975, 54 percent had less than 1,000 population, and 36 percent had less than 500 population.

The postcensal estimates of population and income for those areas are used in a variety of activities, including the allocation of federal funds, public and private planning and decision making, determining the eligibility of a locality for self-government, and scholarly research. The importance of the estimates in determining the allocations made under federal grant programs has been stressed both in the professional literature and in the courts. More than 100 programs make allocations partly or solely on the basis of population estimates (U.S. Congress, 1978). In fiscal 1975, nearly $36 billion was distributed under the 10 largest grant programs that use population and income data to determine allocations (Office of Federal Statistical Policy and Standards, 1978). The general revenue sharing program alone (P.L. 95-512) distributes more than $6 billion a year.

In some instances the postcensal estimates are used to decide whether a place is eligible to receive benefits of one kind or another. The eligibility test may be whether the estimate for the place has exceeded a threshold value. For example, to receive funds under Title I of the Comprehensive Employment and Training Act (CETA) programs (P.L. 93-203), an area must have (or be part of a consortium that has) a population of at least 100,000. To receive funds under Title II of CETA, an area must have a

TABLE 1.1 Local Governments, 1977, and Estimated Population, 1975, by Population in 1970

Population in 1970	Counties			Municipalities and Townships		
	Number, 1977	Percent, 1977	Estimated Population, 1975 (in thousands)	Number, 1977	Percent, 1977	Estimated Population, 1975 (in thousands)
100,000 or more	343	11.3	124,477	194	0.5	62,274
50,000–99,999	336	11.0	23,503	302	0.8	20,890
25,000–49,999	596	19.6	20,976	704	2.0	24,418
10,000–24,999	980	32.2	16,079	1,872	5.2	29,187
5,000–9,999	496	16.3	3,758	2,331	6.5	16,380
1,000–4,999				10,920	30.6	24,225
500–999	291	9.6	897	6,648	18.6	4,734
250–499				5,601	15.7	2,051
0–249				7,112	19.9	946
TOTAL	3,042	100.0	189,691	35,684	100.0	185,105

Note: Because of rounding, population detail may not add to total. The total population of counties (189,691) is lower than the total U.S. population because the 3,042 county governments considered exclude 106 county-type areas that do not possess independently organized county governments. These areas include all of Connecticut, Rhode Island, the District of Columbia, New York City, Philadelphia, and San Francisco, among others.

SOURCES: Bureau of the Census (1978a, Tables B, C, and E) and unpublished data from the Bureau of the Census.

population of at least 50,000. Similarly, to qualify for receiving funds under the Community Development Block Grant program (P.L. 93-383), an area must have a population of at least 50,000. Thresholds can also be in the form of limits: to be eligible to receive funds for state and regional solid waste plans from the Rural Communities Assistance Program, a municipality or county must not be larger in population than 5,000 or 10,000, respectively.

Thresholds apply to activities other than fund allocation. In some states a locality cannot become self-governing unless its population, as determined by a census or postcensal estimate, exceeds a fixed level. Population size also determines how a community is classified by the state government—as a class 1, class 2, class 3, or other kind of city—which delimits the powers, duties, and obligations of the local governmental units. In general, class 1 cities exercise more self-government, have broader taxing powers, and may provide more services than class 2 or class 3 cities. In some states a city's classification also determines such ceilings as the maximum salaries for public officials and the maximum number of establishments permitted to sell liquor.[2]

Many other measures, including employment, unemployment, and birth and death rates, depend implicitly on the postcensal population estimates. Like the postcensal population estimates, these measures are used not only to determine fund allocations but also to identify and analyze problems, to formulate policies to ameliorate the problems, and to evaluate the effects of the adopted policies. These measures are also used in basic scholarly research to formulate and test theories. (Appendix F illustrates the way postcensal population estimates are used in computing official measures of employment and unemployment.)

Planners and decision makers in the private and public sectors rely on the postcensal estimates to evaluate current population trends. Data on these trends are especially useful for heterogeneous regions in which some local areas are gaining and others are losing population. For example, the kinds of plans and decisions that need to be made about education, health, police, and sanitation services differ importantly for growing and declining areas. Particular use is made of the estimates by the more than 200 health systems agencies (HSA's) to develop health plans and review proposed health programs. The postcensal estimates also play a role in the determination of amounts of funds allocated to each HSA under the National Health Planning and Resources Development Act of 1974; those allocations are used to fund promising health programs.

---

[2] Population thresholds for different city classes for the 50 states are given by the Bureau of the Census (1978a).

The private sector depends increasingly on small-area estimates for making decisions about site locations, advertising and promotional campaigns, and market research. Several private companies even specialize in further disaggregating the Census Bureau's small-area estimates into estimates for census tracts within geopolitical boundaries. Small-area data are frequently used for developing estimates for areas other than the standard ones, such as market areas. As more businesses become aware of the existence and utility of small-area estimates, their use is expanding.

1.1c POSSIBLE APPROACHES TO MEET NEEDS FOR POSTCENSAL ESTIMATES

There are many conventions for generating postcensal estimates; they vary in both cost and accuracy. One convention might satisfy the needs of some users but not of others. This section describes seven possible conventions, ranging roughly from least to most expensive. This list is by no means exhaustive but indicates that there are alternatives and that there is a rather wide trade-off between accuracy and cost. For the sake of simplicity these conventions apply to population only; the extension of these conventions to estimates of income would be straightforward.

1. *Use of decennial census counts*[3]   The least costly convention accepts the decennial census counts for the following decade. Thus population counts would be updated only every 10 years. Quite clearly, the estimates would generally become less accurate over the decade since the last census. Yet this convention is currently used for establishing the number of U.S. Representatives to which each state is entitled.

2. *Use of decennial census counts with a rate of change equal to that of the nation or the state*   Convention 1 can be modified to account for estimated growth of the national population over the decade. The simplest way to allocate growth is to assume that every place grows at the same rate as the nation. Such a convention would clearly not differ in effect from convention 1 if a fixed pie were being shared on the basis of population, but some areas would cross thresholds. One could also prepare estimates of the change in population for each state (only 51 estimates) and assume that every place within a state grows at the state rate. This convention would at least allow some regional variation in the distribution of population over the decade. The added cost over convention 1 is minimal.

---

[3] This discussion ignores the fact that not all persons are counted in the census.

3. *Use of decennial census counts updated by natural increase* Since data on births and deaths are available through the vital registration system, the census counts could be updated by the addition of births and the subtraction of deaths since the census date. Such a convention still omits migration, which is known to be a larger component of change than natural increase for many local areas. The added cost over convention 1 is small.

4. *Use of decennial census counts updated by natural increase and estimates of migration* The addition of a migration component considerably enlarges the required sources of data needed to prepare estimates. The United States has no requirement that people report a change of address to a central statistical authority. Hence migration must be estimated by the use of symptomatic indicators such as school enrollment, housing units, etc. or by address information contained in annual Internal Revenue Service (IRS) returns. Even if a migration component is not estimated directly, available methods for estimating population including migrants require data on these symptomatic indicators. This convention is the one now used by the Census Bureau. The added cost of the Bureau's current procedures over convention 1 is perhaps $20 million per decade for six updates over the decade (excluding the cost of updating the geographic coding guide).[4]

5. *Use of decennial census counts augmented by a mid-decade census* If a census enumeration were held every 5 years, one objection to convention 1 would be softened. Perhaps accurate 5-year updates would provide information on change that is sufficiently timely to meet the needs of many users and the requirements of many uses of local area data. The additional cost, however, is probably in excess of $600–$700 million. Of course, a mid-decade census would serve many other uses as well, so the costs should not be attributed entirely to the small-area population and income estimates. Even if the mid-decade "census" did not attempt complete enumeration but was a large-scale sample survey that provided accurate estimates for small areas, the cost would still be great—probably more than $500 million.

---

[4]The geographic coding guide is used by the Census Bureau to assign mailing addresses on Internal Revenue Service individual income tax forms to places of residence. This procedure is important both in making population estimates by the administrative records method and in making postcensal per capita income estimates. The cost of the most recent updating of the coding guide in 1975 was roughly $9 million. (For further discussion, see section 1.2a; Appendix A, sections 2.9, 3.9, and 4.1d; and Appendix K.)

6. *Use of annual censuses*   A complete census might be taken every year. In the view of the Panel the cost would be prohibitive, and public cooperation might wane if information were sought annually.

7. *Creation of a population register*   The analogue of a continuous census is a population register. In effect, each person must register with the local authorities when he or she moves. Such a system is capable, at least in theory, of providing population counts at any point in time. Population registers are maintained in the Netherlands, Finland, Belgium, Norway, Sweden, and Denmark and among certain Indian tribes in the United States. The Panel does not consider this to be a practicable or desirable alternative for the United States.

1.1d CONSIDERATIONS OF ACCURACY

Three factors are important in the production of postcensal estimates of population and income: accuracy, timeliness, and low cost. Timeliness refers to the availability of estimates within a short time after their reference dates. Low cost is usually thought of as a constraint rather than a goal, but actually each of the three goals constrains attempts to satisfy the other two. This section focuses on the accuracy of the estimates.

An estimate is considered accurate if it is close to the value of the parameter (population or per capita income) it is estimating, which is typically unknown. A variety of measures of this closeness or accuracy can be defined. Ideally, an estimating procedure (or estimator) should meet four criteria: (1) low average error, (2) low average *relative* error, (3) few extreme *relative* errors, and (4) absence of bias for subgroups. "Error" is defined here as the difference between the estimate and parameter. "Relative error" is the error expressed as a proportion or percent of the parameter. "Average error" (or "average relative error") refers to the arithmetic mean of the errors (or relative errors) *disregarding sign* (i.e., plus or minus). "Bias" means that an estimating procedure produces estimates that tend to be too high or too low for certain classes of areas.

As is often the case in statistics, it is generally not possible to produce a set of estimates that will minimize all of the above criteria simultaneously, so it is necessary to make choices. Minimization of criterion 2 requires that the relative error in the postcensal estimate be small for a place selected at random. Of course, the actual magnitude of the relative errors will depend on unpredictable circumstances, so at best the relative error for a place can be low with high probability, or the expectation of the relative error can be small. Although the desirability of low average relative error is obvious, this criterion may become controversial if one wishes to give greater importance to some places than to others. A population estimating procedure that minimizes average relative error places greater emphasis on small places than a procedure that minimizes average error: the minimizing of average relative error assigns equal weights to the relative errors of all places, while the minimizing of average error in effect gives larger weights to the relative errors of large areas than to those of small areas.

Criterion 3, few extreme relative errors, means that the relative errors for all places should be approximately the same size. (As was noted above, if errors are random, "same size" means with high probability, in expected value, or in an analogous sense.) Consider a procedure that produces a set of 1,000 estimates with a mean relative error of 4 percent. If all the relative errors are close to 4 percent, the procedure may be very satisfactory, but if the worst 10 percent of cases have an average relative error of 20.2 percent while the best 90 percent of cases have an average relative error of 2.2 percent, the procedure may be very unsatisfactory.

Criterion 4 recognizes that the presence of bias can create political tensions. Estimation procedures rest on demographic and economic assumptions that may not apply to particular classes of areas. For example, as is discussed below (section 1.2a), the administrative records method, which uses information on tax returns to estimate migration rates, depends crucially on an assumption that the proportion of people filing tax returns is the same for migrants into an area, migrants out of an area, and those not moving to or from an area during the given time period.

Criterion 1, low average error, tends to minimize the dollar amounts of misallocated funds under formula grant programs such as general revenue sharing (GRS) because those allocations are often in practice approximately proportional to the fraction of total population residing in an area. Since most of the population live in large areas, emphasis on this criterion implies choosing estimators largely according to their performance in producing good estimates for large areas. This criterion is thus in clear contrast to criterion 2.

In its reports, the Census Bureau indicates primary concern with criteria 2 and 3 (low average relative error and few extreme errors) and some attention to criterion 4 (bias) in its selection of alternative procedures (for example, see Bureau of the Census, 1973b, pp. 2, 10). The Census Bureau is conducting research on the biases in the administrative records method caused by low income-tax filing rates for estimation of interstate migration (Bureau of the Census, 1978c). In evaluating the accuracy of postcensal estimates, the Panel chose to use the same general criteria as the Census Bureau. Thus we considered average relative error, extreme relative error, and bias. The Panel also believes that consideration should be given to the amount of funds that are misallocated because of data error; some general relationships between data error and fund misallocation under the general revenue sharing program are discussed in Appendix E.

The level of accuracy desired for a postcensal estimate of population or income depends on the use for which it is needed. For example, if a threshold is involved, then a high level of accuracy may be of supreme concern. The fact that the population estimate for Trenton, New Jersey, dropped from 101,365 in 1975 to 99,672 in 1976—328 below the threshold of 100,000 required for prime sponsorship for CETA programs—illustrates the importance that a small error could have for a city government (Bureau of the Census, 1977c, 1979). Given the levels of accuracy inherent in the current estimation procedures, the difference of 328 could well have been entirely due to error. On the other hand, if a fixed amount of funds is to be carved up among geographic areas on the basis of population, then only differential error among geographic areas will create disparity between the intent of the legislation and the reality of disbursement. Similarly, private users of such data may tolerate rather large errors, since, for example, the decision to locate a business in an area does not require a precise estimate of the rate at which an area is growing.

1.2 CURRENT ESTIMATION METHODOLOGY

Since the last census in 1970 the Census Bureau has published annual postcensal population estimates for states and counties. Postcensal estimates for subcounty units were first prepared for July 1, 1973, and have been prepared annually beginning with those for July 1, 1975. Per capita income estimates for states, counties, and subcounty units have been produced every year or two since July 1, 1973.[5]

For estimates of postcensal population, the Census Bureau uses essentially two kinds of methods: component methods and regression methods. Component methods first calculate population change, using the number of births minus the number of deaths plus the net number of migrants: the postcensal estimates are the sum of the estimated population change since the last census and the reported population in the last census. In regres-

---

[5] These population and per capita income estimates for states, counties, and subcounty areas are published in the Census Bureau's *Current Population Reports* Series P-25 (see Bureau of the Census, 1974, 1975b, 1979); provisional and revised county estimates appear in *Current Population Reports* Series P-26 (see Bureau of the Census, 1973b).

sion methods, equations are constructed to relate observed population changes to observed changes in other "symptomatic" data that are available and considered relevant. Subsequent observed (postcensal) changes in symptomatic data are then transformed by the equations to yield estimates of postcensal changes in population, which are applied to the reported population in the last census.

Estimation of postcensal population change for subnational areas is at best a complicated process. Numerous data sources are used. Addresses on IRS individual income tax returns for different years are matched to estimate internal migration. Immigration and Naturalization Service records together with passenger statistics (relating to numbers of persons entering and leaving Puerto Rico) form the basis for estimating net immigration from abroad. Data on births and deaths are obtained either from state departments of health or from the National Center for Health Statistics. For many kinds of data the Census Bureau relies on its contacts in the Federal-State Cooperative Program for Local Population Estimates (FSCP). For example, the FSCP members provide to the Census Bureau data on births and deaths from state departments of health; data on populations in institutions and military barracks; school enrollments by county (used in one component method); and administrative data of different kinds, such as numbers of drivers licenses issued, size of the labor force, and numbers of new building permits issued (all used in regression methods of estimation).

For estimates of postcensal per capita income the Census Bureau uses a component method. Income change is viewed as the total of the following: change in wage and salary income, change in social security income, and changes in various other kinds of income. The estimates of changes in income draw upon data from two sources: Bureau of Economic Analysis estimates of components of income for state and counties and IRS individual income tax returns. The Bureau of Economic Analysis uses administrative data from hundreds of sources to make their estimates of components of income (see Coleman, 1978).

For both population and income, errors in the estimates of change can arise both from inappropriateness of assumptions underlying the methods and from errors in the data used. In addition, errors in postcensal estimates of level (rather than of change) can arise from errors in the base-year census data to which the estimates of change are applied. Undercount is a significant source of error in the census counts of population. The Census Bureau estimates that the 1970 census failed to count 5.3 million people, or 2.5 percent of the total population (Bureau of the Census, 1975a). The estimated rates of net undercount vary widely for different subgroups, classified by race, sex, and age. Significantly, the estimated rates also vary for different states (Bureau of the Census, 1977a) and for substate areas.

Sources of error in the census estimates of per capita income include reporting errors, undercoverage bias, and sampling variability. Undercoverage bias arises because unenumerated persons are believed to have different incomes than enumerated persons. Error from sampling variability is substantial for small areas because the sampling rate for the income question in the 1970 census was only 20 percent. For all areas, reporting errors are substantial: it is well known that income is underreported and that the level of underreporting varies sharply by type of income (Bureau of the Census, 1977b; Ono, 1972).

The postcensal estimation methods use current data, and so the estimates appear after their reference dates. The length of delay varies from year to year and by the level of geography of the jurisdiction being estimated. Because so many data sources are used, a delay in arrival of any one set of data can hold up production of the estimates. Several stages of estimates, corresponding to different delays, are published: earliest are "provisional," for counties there are "preliminary," and latest are "revised" or "final." For states, the provisional population estimates appear 8–17 months after the reference date, and revised estimates follow about a year later. For counties, the delays in the population estimates are typically 9–15 months for the provisional, 21 months for the preliminary, and 21–27 months for the revised. The preliminary county estimates are used for determining general revenue sharing allocations. For subcounty areas, only one set of estimates is usually published, roughly 21 months after the reference date. The delays in publication of provisional per capita income estimates for states, counties, and subcounty units are approximately the same as the delays for the subcounty population estimates. The revised per capita income estimates follow about a year later.[6]

The time references for the data do not always correspond to those for the estimates. While the target date for the estimates is July 1, the school enrollment data used to estimate migration pertain to the preceding September or October, and the IRS addresses (also used to estimate migration) pertain to varying dates between the preceding January 1 and April 15. For states and counties, calender year birth and death data are interpolated 6

---

[6] The differences between provisional and revised population estimates are discussed in some detail by the Bureau of the Census (1974, p. 14) for states and in Appendix A for counties. The subcounty population estimates are not generally revised (except for the 1973 estimates, which were revised because of changes in the geographic coding procedures of the Census Bureau). Revisions in per capita income estimates result from changes in data rather than changes in procedure.

months to attain exact time correspondence. For subcounty units, because of the lack of available data for many places, the estimates of net natural increase are not interpolated but refer to the preceding calender year and so are 6 months out of synchronization. However, the numbers of births and deaths for subcounty units are scaled to sum to county totals.

1.2a POPULATION ESTIMATION

This section summarizes the methodology used by the Census Bureau to prepare its postcensal population estimates (see Appendix A below for further details). We should note at the beginning that the population of an area is the number of persons whose place of usual residence is in the area; it includes both legal residents and those not legally permitted to reside in the United States.

The estimates for different geographic levels are produced in a hierarchical manner. National estimates are produced first. Then state estimates are produced and "controlled" to the national estimate: that is, the state estimates are scaled to sum to the previously derived national estimate. County and subcounty estimates are controlled to state and county totals, respectively.

The Census Bureau uses several methods to produce postcensal population estimates. To estimate total U.S. population, a component method is used to account for births, deaths, and net immigration. State and county population estimates are derived as averages of the results of three procedures: a component method, component method II (CM II); an administrative records method (AR); and a ratio-correlation method (RC). Generally, subcounty estimates are derived from the AR method alone.

The CM II and AR are component methods that analyze population change by estimating the demographic facts of birth, death, and migration. Ideally, information about components of population change could be recorded from time to time as events of birth, death, and changes of residence occur. Updating the population level of an area would then be a simple matter of adding to the population at some initial time the components of population change during the period up to the reference date of interest. Such an ideal situation is far from the case for the United States. People changing their place of residence are not required to report to a central agency. Births and deaths are registered individually by place of occurrence (rather than by place of residence); the aggregate statistics are tabulated by place of residence for all counties and for all subcounty jurisdictions with (1970) population of more than 10,000 but not generally for subcounty jurisdictions with population of less than 10,000.

Less information is available on internal migration than on births and deaths. There are no directly relevant administrative records, so symptomatic data are used. CM II uses changes in school enrollment to derive an estimate of net migration of population under age 65 for a state or a county; AR uses the federal individual income tax returns, matching for changes of address, to estimate net internal migration of a state or county population under age 65 and of a subcounty population in general. That is, AR matches individual tax returns for successive periods and determines for each area the numbers of inmigrants, outmigrants, and nonmigrants represented by the returns (taxpayers and their dependents). From the difference between the inmigration and outmigration rates of taxpayers and dependents, a net migration rate is calculated and applied to a base population figure, yielding an estimate of net internal migration. An important part of this process is determining to which of the 39,000 geographic areas of residence the tax returns should be assigned. The mailing address is often insufficient for determining place of residence, and questions on residence were asked on the 1972 and 1975 tax returns. The information from these questions is used to construct geographic coding guides to assign mailing addresses to places of residence. (For further discussion, see Appendix A, section 4.1d.)

The AR method estimates immigration and emigration separately from internal migration. Although alien immigration is legally controlled by the Immigration and Naturalization Service, the number of aliens who enter and reside in the country without a legal status has been a statistical as well as administrative problem. Finally, emigration of many U.S. residents to other countries may never be reflected in aggregated statistical or administrative records.

Various categories of people are treated differently in component methods. People living in group quarters, such as college students, people in institutions, and people in military barracks, are treated separately because these special populations are obviously not subject to the same "risk" of birth, death, or migration as the rest of the population. In addition, whenever appropriate and feasible, estimates of changes in birth, death, and migration are also differentiated by age, sex, and race. For example, at state and county levels, the elderly population of age 65 and above are treated as a special population, and changes in the number of elderly people are estimated on the basis of Medicare data.

At the subcounty level, there are complications involved in estimating births and deaths because data on births and deaths are generally not available for subcounty places with less than 10,000 population (representing more than 90 percent of subcounty units); estimation of these components of population change must be indirect (see Appendix A, section 4.1b for details).

The ratio-correlation method (RC) is a regression method rather than a component method. Regression methods are based on the fitting of a relationship (usually by least-squares regression) between the population change of an area and changes in symptomatic variables. The relationship, or model, is fitted on the basis of information available for the two preceding decennial censuses. The relationship is then used to generate postcensal population estimates when current data are substituted for the symptomatic variables (see Appendix A, section 2.5 for details).

For state population estimates, some of the symptomatic variables used are the number of students enrolled in elementary schools, of federal income tax returns, of registered passenger cars, and of people in the work force. At the county level, other variables are included in the equation if the data are available for all counties in the state. Another difference in the application of RC to estimates of the state and county population involves people living in group quarters. At the state level, RC is used only for estimates of non-group quarters population under age 65, while the rest of the state population is estimated as in CM II. At the county level, RC is used to estimate the whole non-group quarters population.

Occasionally, estimates produced by other methods are included in the Census Bureau's average estimate. For example, a drivers license address change method (DLAC) is used by California to estimate county populations. DLAC is a component method that uses drivers license address changes for estimation of net migration. In Florida a housing unit method (HUM) was used for county estimates in 1975. HUM estimates the non-group quarters population by the product of the estimated average number of persons per household and the estimated number of occupied housing units. These estimates usually are produced not by the Census Bureau but by participants in the Federal-State Cooperative Program for Local Population Estimates (see section 1.2d). These estimates are more often available for counties than for subcounty areas, but some state agencies also prepare subcounty estimates. Since the Census Bureau requires that estimates within a state be the product of a uniform methodology, the estimates from these other methods are taken into account only if they are provided for all counties or subcounty areas within a state.

Special censuses for county and subcounty jurisdictions may be undertaken by the Census Bureau on the authorization of the appropriate local government.[7] The local government pays the necessary expenses and pro-

---

[7] The Census Bureau was also required by the Voting Rights Act of 1965 (42 U.S.C. § 1973 aa-5, as amended by P.L. 94-73) to conduct special censuses for jurisdictions meeting certain criteria in order to determine whether more than 50 percent of the nonwhite persons in the jurisdiction were registered to vote. In the vast majority of cases, however, a special census is taken by the Census Bureau only if a local government requests it.

vides office space and equipment; the Census Bureau provides the personnel. The content of a special census is ordinarily limited to questions on relationship to head of household, age, race, and sex, but additional questions may be included at the request and expense of the sponsor. For areas with less than 50,000 population the costs range from about $0.50 to $1.00 per person; for larger areas the costs are higher. In some states, notably Oregon, Washington, and California, special censuses are conducted predominantly by state agencies.

To combine different estimates, the Bureau first controls to higher level totals (e.g., all county estimates must sum to the state estimate) and then averages the different estimates, assigning equal weights to each. When the results of a special census are available for a county or subcounty area, they are used instead of the various postcensal estimates. In those situations the adjustment of county (subcounty) estimates to sum to the state (county) estimate follows a complicated procedure, sometimes called "rake/float" (see Appendix A, section 4.2 for details).

State population estimates, whether provisional or revised, are derived as equally weighted averages of the estimates from the component method II, the ratio-correlation method, and the administrative records method. The methods used to produce county population estimates vary, depending on whether the estimates are provisional, preliminary, or revised. Generally (for exceptions and more details, see Appendix A, section 3.1), revised county estimates are derived as equally weighted averages of the estimates from CM II, AR, and RC. Preliminary county estimates (used for general revenue sharing) are generally obtained as the sum of the previous year's revised estimate plus the average of two estimates of change during the year, one derived from CM II and the other from AR.[8] Provisional county estimates are obtained as the sum of the previous year's revised

---

[8] The weighting for year t can be represented as follows (for simplicity we ignore the inclusion of locally prepared estimates):

$$\text{preliminary estimate } (t) = \text{revised estimate } (t - 1)$$
$$+ \tfrac{1}{2}[\text{CM II}(t) - \text{CM II}(t - 1)]$$
$$+ \tfrac{1}{2}[\text{AR}(t) - \text{AR}(t - 1)]$$
$$= \tfrac{1}{3}\text{RC}(t - 1)$$
$$- \tfrac{1}{6}\text{CM II}(t - 1)$$
$$- \tfrac{1}{6}\text{AR}(t - 1)$$
$$+ \tfrac{1}{2}\text{CM II}(t)$$
$$+ \tfrac{1}{2}\text{AR}(t).$$

To derive the second equality, note that revised estimate $(t - 1) = \tfrac{1}{3}[\text{RC}(t - 1) + \text{CM II}(t - 1) + \text{AR}(t - 1)]$. The RC estimates for year t are not used for deriving the provisional estimate for year t because they are not available at the time the provisional estimates are produced; they are used for the revised estimates.

estimate plus an estimate of change during the year, where the change is estimated either by CM II alone or by the average of CM II and HUM.

Table 1.2 summarizes the application of the methods for providing different estimates.

1.2b PER CAPITA INCOME ESTIMATION

The Census Bureau's definition of per capita income (PCI) is the average amount of income received per person during the preceding calendar year by all persons residing within a defined political jurisdiction as of the estimate date. (The methodology is summarized in Appendix B.) The per capita income estimates are based on the concept of money income. The Bureau of the Census defines total money income as the sum of (1) wages and salary income, (2) net farm self-employment income, (3) net nonfarm self-employment income, (4) social security and railroad retirement income, (5) public assistance income, and (6) all "other" sources of money income including interest, dividends, pensions, unemployment insurance, alimony, veterans' payments, etc. The total money income represents income received prior to personal income tax, union dues, or any other deductions.

The PCI estimates for different geographic levels are, like population estimates, produced in a hierarchical manner. State estimates are produced, then county estimates, and last, subcounty estimates. County (and subcounty) estimates are controlled to the state (and county) estimates in several ways. For example, the estimates of wages and salary income for

TABLE 1.2 Methods Used by the Census Bureau for Making Substate Population Estimates

Method	County			Subcounty
	Provisional	Preliminary	Revised	
CM II	X	X	X	
AR		X	X	X
RC			X	

Note: When more than one method is listed, the estimates are averaged. The state and county provisional and revised estimates are derived by adding to a previous revised estimate (or census count) the change calculated by the method or average of methods used. For counties and subcounty areas in some states, additional methods are used by state agencies participating in the FSCP, and the resulting estimates are averaged by the Census Bureau with the Bureau's estimate(s).

all counties in a state are constrained to sum to the wage and salary income for the whole state.

The general technique for estimating state PCI is to update the 1970 census estimate of total money income to account for changes in income and then to divide by the estimated postcensal population. Money income is updated using administrative data from two major sources, the Internal Revenue Service (IRS) and the Bureau of Economic Analysis (BEA). The BEA develops its data from several hundred data sources, including many kinds of administrative records (Coleman, 1978). The Census Bureau estimates money income by component: wage and salary information comes from IRS data on gross income reported on individual income tax returns, and the remaining five components of money income are updated on the basis of BEA estimates of personal income.

Personal income and total money income are different concepts of income, and the BEA data must be adjusted. For example, the BEA data refer to income where produced (place of work) rather than income where received. Adjustments are performed to convert the BEA data to a place of residence basis, as used by the money income concept. These adjustments can be substantial for areas where many workers commute. Also, the BEA data include estimates of in-kind income, such as imputed rents and food produced for home consumption. In-kind income is not a component of money income and must be excluded from the BEA data before it can be used to update money income. Other adjustments are also made in the BEA data to attain compatibility with the money income concept. (See Appendix B for further discussion of the role of BEA's personal income estimates.)

County PCI updates are developed in generally the same manner, except that the Census Bureau updates county wages and salary income intact as a per capita figure, on the basis of changes in IRS data on gross income per exemption on the individual income tax returns. Another difference between the methodology for county PCI and state PCI centers on the estimation of farm self-employment income. Farm income is notoriously volatile, capable of sharp year-to-year changes, which may be understated or overstated by the data used to estimate them. To prevent unwarranted sharp fluctuations in its estimates of county farm income, the Census Bureau uses two farm income estimates and constrains the rates of changes in these estimates. (See Appendix B for details.)

Subcounty PCI is estimated roughly the same way as county PCI. Special considerations are necessary because BEA data are not available for subcounty areas. To update subcounty PCI, the Census Bureau decomposes money income into two parts: transfer income (TI) and adjusted gross income (AGI). The TI is composed of social security income, public assistance income, and some parts of "other" income, such as unemployment compensation and veterans' payments. The AGI is the rest of money income. The Bureau estimates TI by assuming that the rate of change in subcounty TI is the same as the rate of change in county TI. Change in AGI is estimated from the income reported on income tax returns. The rates of change are applied to base period estimates to yield estimates of the level of postcensal per capita income.

Because the postcensal PCI estimates are obtained by applying rates of postcensal change to base period estimates, weaknesses in the 1970 census estimates affect the postcensal estimates. The 1970 census estimates of PCI (calendar year 1969) were based on 20-percent samples, and so the PCI estimates for the smallest places are subject to large sampling variance. Hence the Census Bureau did not attempt to estimate directly the 1972 PCI for places with 1970 population under 500 but used the county PCI estimate for these places. Using recently developed statistical techniques (Fay and Herriot (1979); see also Appendix J), the Bureau was subsequently able to revise its estimates of 1969 PCI and produce PCI estimates for those small places.

Numerous other substitutions, constraints, and edits to the data are used to adjust for weaknesses in the data. For example, to compensate for conceptual differences between BEA and Census Bureau income concepts, the county farm income estimates are constrained to fall within 80–120 percent of an alternative estimate.[9] This constraint affects about one-quarter to one-third of the counties. Many of the substitutions, edits, and constraints for the subcounty data are designed to protect against errors in attributing IRS tax returns to the wrong geographic area. The problem of assigning the correct geographic area of residence for the filer of a tax return is significant for the per capita income estimates as well as for the population estimates. Other constraints restrict estimates of relative change for subcounty units to be close to the relative change for the county as a whole. These constraints damp changes but yield more plausible and presumably more reliable subcounty estimates. Complicated controls are also employed to force subcounty estimates for classes to sum to county totals. (For further discussion, see Appendix B.)

1.2c REVIEW OF THE ESTIMATES: CHALLENGE PROCEDURES

An important part of the estimation program is the process of local review. Before the population estimates are published, the Census Bureau

[9]This alternative estimate is the "gross change" farm income estimate; see the section on county updates in Appendix B.

sends to each local area the following: its population estimate, a brief summary of the methodology used to obtain the estimate, a description of alternative estimating techniques and data sources, and a general outline of procedures for reviewing and challenging the Bureau's estimates. (A copy of the outline of procedures for the 1977 population estimates appears in Appendix D.) Each local area responds if it thinks the estimate should be changed. The Bureau keeps a log of all these challenges and subjects each to a detailed review. This review includes examination of data provided by the locality in support of its challenge and also a second careful check of the data used by the Census Bureau to derive its estimate. In some cases the Bureau revises its estimate. More often the local authorities do not provide sufficient data to support their challenges, and the Census Bureau declines to revise its estimate. In the latter case, informal discussions take place between officials of the local area and of the Census Bureau to try to resolve the challenge. If these informal discussions fail, a state or unit of local government may request a formal hearing.

The local review process for per capita income estimates is slightly different. These estimates are sent for review not to each local area but rather to members of the Bureau of Economic Analysis's "user group." The group comprises several people from each state who review the BEA personal income estimates for counties and the census per capita income estimates for counties and subcounty areas in their respective state. They forward comments on the estimates directly to the Census Bureau. The local officials themselves have an opportunity to review their estimates when the Office of Revenue Sharing gives them advance notice about the data elements on which their GRS allocations will be based. At this point the local areas may challenge the Office of Revenue Sharing or the Census Bureau. In either case, the Census Bureau will review its estimate and possibly revise it. Local areas usually have few data with which to support their challenges. In scrutinizing the derivation of the estimate, the Bureau may nevertheless discover an anomaly and revise its per capita income estimate. If the Bureau fails to revise the per capita income estimate to the local government's satisfaction, that government may request a formal hearing.

The Bureau has only recently established the procedure for a formal hearing.[10] The major provisions of the procedure (1) require that an informal challenge be filed within 180 days after the release of the estimates, (2) require that informal review be completed before a formal hearing is allowed, (3) provide for the appointment of a hearing officer (employed by

---

[10] A set of rules for the hearings appears in *Federal Register* (1979, pp. 20,646-20,649).

the Census Bureau but not involved in the preparation of the estimates) to receive evidence under oath, (4) allow for the cross-examination of both parties in the proceedings and of any witnesses, and (5) set time limits for the initiation and completion of the formal challenge proceedings.

In the short period since the provisions for a formal hearing were established, none has been requested. Neither have there been any challenges in court to the Census Bureau's postcensal estimates of population and income, despite the fact that the complaints and informal challenges to the Bureau's estimates are numerous—roughly 50–100 per year for the income estimates and several thousand per year for the population estimates.

1.2d FEDERAL-STATE COOPERATIVE PROGRAM

The Census Bureau initiated the Federal-State Cooperative Program for Local Population Estimates (FSCP) in 1967. The basic goal of the FSCP was to provide high-quality, consistent series of county population estimates with comparability from area to area. The participants in the FSCP are officially designated state agencies.

The FSCP plays several roles in the Census Bureau's postcensal estimation program. As was mentioned earlier, the FSCP contacts provide to the Census Bureau many kinds of data used to make the postcensal population estimates. The state agencies also provide review and comment on the Census Bureau's preliminary county estimates. This working relationship is beneficial to the Census Bureau because the FSCP members have easier access to these data and are in a better position to evaluate the data and correct some kinds of errors. The FSCP members are also better situated to discover new or additional data series that can be used in producing population estimates. The state agencies in the FSCP may also produce population estimates that the Census Bureau uses in making its own estimates.

The 49 states now participating in the program (all but Massachusetts) have designated state agencies to deal with the Census Bureau. While early efforts were limited to estimates for counties, several members of the FSCP now produce subcounty estimates as well. When the Census Bureau uses the FSCP estimates, it first controls them to totals and then averages them with the Bureau's own estimates.

At present, the FSCP operates with very modest resources. The Census Bureau has put considerable energy of skilled professionals into methodological research, experimentation, and evaluations and into technical guidance for the states, but unlike other federal-state cooperative programs (such as the employment, hours, and earnings system of the Bureau of Labor Statistics, the cooperative system of the National Center for Health Statistics, and the crop reporting system of the U.S. Department of Agriculture) the population estimates program provides no financial support or payments to the state agencies. On the basis of available studies of state demographic activities, it appears that most of the agencies participating in the FSCP are underfinanced and short of qualified personnel (see Rosenberg and Myers, 1977). A consequence is that the system must forego the improvement that might result from stronger state programs.

1.3 FINDINGS AND CONCLUSIONS

1.3a SUMMARY

The Panel finds that the methodology of the three population estimation procedures used by the Census Bureau is generally sound.[11] The Panel also commends the Bureau for attempting to measure the error of its estimates and for publishing the results.

Despite the basic soundness of the estimation methods, however, they result in estimates that are directionally biased for some categories of local areas. They also result in large random errors for other areas, especially small subcounty areas (those with less than 2,500 population) and subcounty areas of moderate size (those with up to 25,000 population) undergoing rapid growth or decline in population. For example, for subcounty areas for which special censuses were taken in 1975, the average error in estimates of total population was 23 percent for areas with less than 500 population and 10 percent for areas with 1,000-2,499 population. (More than one-third of the subcounty areas eligible for GRS funds had less than 500 population in 1975.) The average error for areas of very rapid population growth—defined here as an increase in population of 50 percent or more between 1970 and 1975—varied from 27 percent for those with less than 500 population to 19 percent for places with 10,000-24,999 population and then dropped sharply to 7 percent for places with 25,000 or more population (see Table 2.8).

The Panel has several proposals for technical modifications of the estimation procedures, which may improve the accuracy of the estimates to some degree, especially for counties and large cities. However, the

---

[11] For county estimates the Bureau uses unweighted averages of three methods: component method II, ratio-correlation, and administrative records. For subcounty estimates, available data permit the use of only the administrative records method.

Panel knows of no feasible procedure within the limits of present data sources that would significantly reduce the errors in population estimates for small subcounty areas. Accurate estimates for small areas cannot be developed unless data collection is increased enormously, by such means as more frequent censuses or a compulsory registration system.[12]

The task of estimating per capita income for small areas is even more formidable than that of estimating population. Because of severe data problems, the estimates of postcensal per capita income are less accurate than those for population. The Panel does not have any recommendations for improving the methodology for estimates of per capita income, nor do we know of alternative data sources that might produce substantially more accurate estimates at acceptable cost.

In our opinion the Census Bureau is in an unnecessarily difficult situation. It is required to defend (to the last digit) population estimates that its own analyses have shown may have relative errors of 25 percent or more. The mushrooming amount of legislation that authorizes distribution of funds on the basis of population or income estimates for small areas gives increased incentive for officials from these areas to challenge the Bureau's figures, and an increasing share of the Bureau's energies must be devoted to these challenges.

In evaluating the Census Bureau's program for postcensal estimates, the Panel assessed the accuracy of the estimates and examined the logic of the methodology used to produce the estimates. In addition, the Panel tried to identify some of the key decisions made when the statistical methodology was developed.

The available information indicates that the postcensal population estimates are most accurate for areas with large populations and moderate rates of population growth or decline. The relative error[13] of the estimates increases as the population size of the area decreases and also as the percent change in population (growth or decline) increases. In general, the estimates for counties are quite accurate: the average relative error was 3.9 percent for 133 counties in which special censuses were taken from 1974 to 1976. The population estimates for subcounty areas with small populations were highly inaccurate: the average relative error was 23 percent for subcounty areas with less than 500 population and 10 percent for

---

[12] The recent report of the National Commission on Employment and Unemployment Statistics (1979) arrived at a similar finding about labor force statistics: that there is no way, at reasonable cost, to produce accurate employment and unemployment statistics on a current basis for thousands of local areas.

[13] The measure of average relative error used was the arithmetic mean of the percent differences (disregarding sign) between the population estimate and the special census count, generally referred to as the "average percent difference."

areas with 500 to 2,499 population. (These results are based on percent differences between estimates and special census counts for 799 subcounty areas in which special censuses were taken during 1975.) As we noted above, more than one-third (36 percent) of the more than 35,000 subcounty areas for which the Census Bureau prepares estimates of population and per capita income had less than 500 population in 1975. (See section 1.3b for more discussion and section 2.2c for further details.)

Estimation of postcensal per capita money income is an especially difficult task. As was noted above, because of severe data problems the postcensal estimates of per capita income are less accurate than those of population. The limited evidence available indicates that accurate income estimates cannot be produced even for subcounty areas with populations from 10,000 to 20,000. No evaluation data were available for county estimates. (See section 1.3d for more discussion and section 2.3 for details.)

The methodology of the Census Bureau's per capita income estimates is well designed, but problems exist because of data limitations and because of the conceptual basis for the estimates (see section 5.1e). The estimation procedure draws heavily on the county personal income estimates of the Bureau of Economic Analysis. Personal income and money income have different conceptual bases. Hence complicated adjustments of questionable accuracy must be applied to the personal income data; the problems are particularly severe for areas in which farm income is a substantial part of total income. These areas include many of the smaller subcounty areas and counties.

1.3b POPULATION ESTIMATES

In evaluating the quality of the estimates of population the Panel has examined both the logic and the accuracy of the techniques used to produce the estimates. In examining the logic we considered whether the procedures made sense from the standpoint of demographic and statistical theory. To study the accuracy of the estimates, we relied primarily on comparisons of postcensal estimates with the results of special censuses taken during the 1970s.

Special censuses are censuses conducted for municipalities, townships, or counties within a state that are not part of a national effort. Places that have special censuses are usually self-selected; they are not a random sample of all places. They choose to have a census, and they pay for it. Areas are more likely to have special censuses if they expect a special census to document a substantial increase in population. Such places tend to have higher-than-average growth rates, but it is known that postcensal population estimation methods perform worse for those places than for slowly growing places. A way to avoid the bias that exists in the selection of places for special censuses would be for the Bureau to underwrite the cost of special censuses for a probability sample of local areas as it did in 1973 for 86 areas. Such a sample would be the strongest way to test the accuracy of the estimates, but it would be prohibitively expensive to do for a sufficiently large number of areas to provide reliable estimates of error for the full range of population-size and rate-of-growth subgroups of areas.

1.3b(1) *County Estimates*

Estimates of county population are, on the average, quite accurate. For 133 counties receiving special censuses between January 1, 1974, and December 31, 1976, the average difference (disregarding sign) between the postcensal estimates and adjusted special census counts was 3.9 percent.[14] The accuracy of the estimates varied with the population size of the county, and with the percent change in population size, as follows (data are from Table 2.1).

		1970 Population				
	Total	Under 1,000	1,000– 4,999	5,000– 24,999	25,000– 99,999	100,000 or More
Average percent difference	3.9	7.1	5.2	3.6	2.9	1.4
Number of counties	133	24	23	32	22	32

	Percent Change in Population, April 1, 1970, to July 1, Year of Special Census					
	−5.0 or More	−0.0 to −4.9	+0.0 to +4.9	+5.0 to +14.9	+15.0 to +24.9	+25.0 or More
Average percent difference	5.1	4.0	3.3	2.4	3.6	6.8
Number of counties	11	16	25	42	17	22

The populations of large counties are estimated more accurately than those of small counties; those of slowly growing counties are estimated more accurately than those of rapidly growing or declining counties. For

---

[14] The percentage base is the special census adjusted (by linear interpolation or extrapolation) to refer to the nearest July 1, which is the date for the postcensal estimate.

example, the average percent difference between the estimate and the special census count decreased monotonically from 7.1 for counties with 1970 population of less than 1,000 to 1.4 for counties with 1970 population of more than 100,000. The average percent difference was only 2.4 percent for counties growing between 5 and 15 percent but was 6.8 percent for counties growing by 25 percent or more.

There also is evidence of bias in the county estimates: they tend to underestimate the change in population since the last census, both when population is increasing and when it is declining. For example, the estimates were too high for 8 of 11 counties that declined in population by 5 percent or more, while estimates were too low for 32 of 39 counties that had grown by 15 percent or more (see Table 2.2).

1.3b(2) *Subcounty Estimates*

Estimates of population for subcounty areas are less accurate than estimates for counties of the same size. For example, counties with 1,000 to 4,999 population had an average error of 5.2 percent; subcounty areas of the same size had an average error of 8.8 percent (see Tables 2.1 and 2.9).

The population estimates of subcounty areas in 1975 were quite accurate for areas with large populations but were increasingly inaccurate as population size decreased.[15] For example, the average percent difference between 1975 population estimates and comparable 1975 special census counts was only 2.6 to 2.7 percent for areas with population of 25,000 or more in 1970 but increased to more than 25 percent for areas that had less than 250 population in 1970 (see Table 2.7).

The accuracy of the estimates also varied greatly by the rate of population change between 1970 and 1975. Areas with relatively stable populations—less than 5 percent growth or decline—had an average error of 6 percent; areas that grew by 50 percent or more or that declined by at least 10 percent had average errors of more than 20 percent.

Estimates for areas that were both small and had experienced rapid growth or decline were most inaccurate. For example, for subcounty areas with less than 500 population that declined in population by 10 percent or more between 1970 and 1975 the average error was 43 percent; for areas of the same size that grew by 50 percent or more the average error was 28 percent (see Table 2.8).

In general, very small areas (those with less than 500 population) had

[15] This evaluation of the accuracy of postcensal estimates of population for subcounty areas was based on comparisons of 1975 population estimates with special census counts for 799 subcounty areas that were taken in 1975.

large errors regardless of the rate of change in population. Only large areas (25,000 or more population) had relatively accurate estimates regardless of the rate of change in population; average percent differences in this population-size group varied from 2.4 percent for areas that changed by less than 10 percent (growth or decline) to 6.6 percent for those that grew by 50 percent or more.

There is also strong evidence of bias in the subcounty population estimation methods: they consistently tended to underestimate the population of growing areas and to overestimate the population of declining areas. In our comparisons, for example, more than 84 percent of the estimates for areas that declined in population by 10 percent or more between 1970 and 1975 were overestimates, and more than 91 percent of the estimates for areas that grew by 50 percent or more were underestimates.

The low levels of accuracy of the estimates for small areas, and for areas undergoing rapid growth or decline, are also evident in measures of "extreme error." Over half of the subcounty areas with less than 500 population had relative errors of 15 percent or more.

The measures of error discussed thus far are relative errors in the estimates of total population of subcounty areas. But the estimation methods are designed to measure change in population since the last census (since the total population estimates are obtained by adding the estimated change to the previous census counts). Moreover, the usefulness of the estimates as updates for the purpose of allocating general revenue sharing funds between regular censuses depends on the accuracy of the estimated changes in population. Hence the Panel also calculated measures of the relative error in the estimates of change in population since the last census.[16]

The errors based on change in population were, for the most part, many times larger than comparable errors in the estimates of total population, and the pattern of error was substantially altered. Subcounty areas subject to little growth or decline had the largest relative errors based on change in population, whereas the fast-growing areas had much smaller errors. From this perspective the greater accuracy of estimates of total population for slowly or moderately changing areas as compared with areas of rapid growth or decline can be explained by the fact that their change in population from 1970 to 1975 was a smaller proportion of their total population in 1975 than was the case for areas undergoing more rapid growth or decline.

---

[16] Relative errors based on change in population were calculated as averages of the percent differences between estimated change in population (1975 estimate minus 1970 census count) and enumerated change in population (1975 adjusted census count minus 1970 census count); see section 2.2c for further discussion.

The relative errors based on change in population are exceedingly large. For example, the average difference of 6.4 percent (based on total population) for areas with 1,000 to 2,499 population that grew by 5-9 percent between 1970 and 1975 represents an average percent difference of 100 percent between the estimated and enumerated change in population from 1970 to 1975. Similarly, the average difference of 10.7 percent (based on total population) for areas with 1,000 to 2,499 population that grew by 10-24.9 percent from 1970 to 1975 represents an average difference of 63 percent (based on change in population); (see Table 2.12). More importantly, in 20 of the 118 subcounty areas with 1,000 to 2,499 population, the estimated change in population was in the wrong direction—increase instead of decrease, or *vice versa*. In an additional 9 areas the change in population was overestimated by more than 100 percent.

The large relative errors in estimated change in population raise questions about the advisability of attempting to update population data for purposes of allocating funds to small areas. It seems quite possible that the estimated postcensal change in population for a substantial proportion of the subcounty areas below some as yet unspecified threshold may be in the wrong direction or may have average errors of more than 100 percent. This possibility should be carefully checked in the test of the estimation methods against the 1980 census results.

1.3c METHODOLOGICAL DECISIONS FOR POPULATION ESTIMATION

The Panel has examined several decisions the Census Bureau made when it developed its estimation methodology. These decisions concern (1) the selection of estimation methods, (2) the uniformity of estimation procedures, and (3) the averaging of estimates. Since only one method was generally used for making subcounty estimates, points 2 and 3 currently pertain more to the county and state estimates than to the subcounty estimates. Since additional methods may be used for large subcounty areas in the future, however (if, for example, the uniformity criterion is relaxed), averaging and uniformity considerations may also become important for subcounty estimates.

1.3c(1) *Selection of Methods*

The Census Bureau has based its subcounty population estimates solely on the administrative records method.[17] In designing an estimation method-

---

[17] As was mentioned above, some state agencies in the FSCP also produce subcounty estimates which the Census Bureau averages with its own.

ology the Bureau needed methods that could be used to produce estimates for all the 36,000 subcounty jurisdictions participating in general revenue sharing. The input data (IRS individual income tax records) for the AR method are available for all subcounty units, but data needed for the other methods used to produce county estimates are not available for all GRS jurisdictions below the county level.

The AR method was developed after 1970. The first tests of the method were performed for 16 counties and 8 subcounty areas with populations of more than 50,000 (Bureau of the Census, 1975b; Zitter, 1972; Zitter and Word, 1973). Later the AR estimates were tested against special censuses taken in 1973. These tests (Bureau of the Census, 1975b, Tables D–G) included comparison of the AR estimates with results of special censuses for a probability sample of 86 areas with population of less than 20,000 and for 165 areas where special censuses were conducted by the Census Bureau at the request and expense of the locality (these were *not* a random sample of subcounty areas). The Bureau's decision to use the AR method to make subcounty population estimates was based partly on these limited tests and partly on *a priori* considerations relating to the extensive coverage of the IRS data and the lack of workable alternatives (Bureau of the Census, 1975b). While more testing would have been desirable, it is the view of the Panel that the Census Bureau did as much as might reasonably be expected, given the pressures of time after the general revenue sharing legislation was drafted. The Panel believes, however, that more testing is called for when decisions about the choice of future estimation methods must be made.

For county population estimates, several methods are available. Prior to the 1970s the Census Bureau had traditionally relied primarily on four types of estimates: ratio-correlation, component method II, composite, and vital rates. Tests for 2,586 county estimates against the 1970 census (Bureau of the Census, 1973b, Table C) indicated that the Bureau's RC method was clearly superior to the other three methods but that there were circumstances in which judicious averaging of CM II or composite method estimates with RC estimates produced results that were better than those obtained with RC alone (Bureau of the Census, 1973b, Table D).

On the basis of these tests the Bureau decided to use its ratio-correlation method (RC) and component method II (CM II) in its county estimation methodology. The Bureau dropped the composite method for use in the 1970s, despite its good test performance. The reasons for dropping the composite method are not reported in publications, but Bureau staff have indicated that it was done for the same reasons that births were dropped as a predictor variable in the RC estimation of state populations (see Bureau of the Census, 1974), namely, because of changes in laws permitting abortions (in the early 1970s), which were followed by unusually large drops in the birth rates for many large metropolitan counties. With the composite method, which uses changes in the birth rate in estimating changes in the population aged 18 to 44, these drops would have implied substantial drops in the population of these counties. Since the CM II, RC and AR methods did not indicate comparable drops in population, the Bureau concluded that the composite method might not give reliable estimates of total population.

1.3c(2) *Uniformity of Methodology*

The Bureau currently uses a uniform methodology to produce its estimates. The same methods are used for all counties in a state, regardless of size of area, rate of growth, or unusual age or other composition factor. Similarly, a uniform method is used for all subcounty areas. If data are available for some but not all counties or subcounty areas within a state, the Census Bureau does not use those data. The Census Bureau imposed the uniformity constraint on itself because it thought the estimation procedure would be more defensible against challenges and because restriction to uniform procedures makes the estimation program more manageable. The Panel believes, however, that the accuracy of the estimates might be improved if the uniformity constraint is relaxed.

1.3c(3) *Averaging of Estimates*

To estimate county and state populations, the Bureau averages the results of different methods. As was described above (section 1.2a), revised county estimates are generally obtained as the equally weighted average of the CM II, AR, and RC methods. Preliminary county estimates (used for general revenue sharing) are generally obtained as the sum of the previous year's revised estimate plus the equally weighted average of two estimates of change during the year, one derived from CM II and the other from the AR. The different methods are not equally accurate—CM II is less accurate than the AR or RC methods (see section 2.2)—and the estimates can be improved if unequal weightings are used. For example, use of unequally weighted averages reduced the average difference between estimates and special census counts for 130 counties from 4.3 to 4.0 percent (see Table 5.2).

1.3d PER CAPITA INCOME ESTIMATES

Evidence about the accuracy of the per capita income (PCI) estimates is derived solely from tests based on a random sample of 86 special censuses (in areas with less than 20,000 population) conducted and paid for by the

Census Bureau. These tests indicate a large difference between 1973 postcensal estimates of PCI and PCI measures from special censuses taken in 1973. The average difference (without regard to sign) for all places with population of 1,000 to 20,000 was 10 percent of the special census PCI. For places with a 1970 population of less than 500 the average difference was 28 percent of the special census PCI. For places with populations between 500 and 999 the average difference was 17 percent (see Table 2.13, column 4). After revisions to their methodology for places with population of less than 1,000, the Census Bureau reduced the estimated levels of error for these small places by a few percentage points (see section 2.3).

The available data are not adequate to draw any conclusions about how the error levels decrease as population size of the areas increases. Theoretically, the Census Bureau should be able to make more accurate PCI estimates for larger areas than for smaller ones. The base figure with which the Bureau must work is more accurate for larger areas, and subsequent adjustments should be more accurate. Adjustments to the 1970 census data are easier for metropolitan areas because wages and salaries are such a large component of income and current estimates of these are available. Estimating wages and salaries is a relatively simple task compared to estimating proprietors' incomes, that is, the net business earnings of owners of unincorporated enterprises.

As was noted above, for purposes of allocating general revenue sharing funds it is *changes* in population and per capita income since the last census that produce changes in the allocation of GRS funds on postcensal dates. Hence it is the accuracy of the estimates of change in per capita income since the last census that is important in evaluating the use of the income estimates for updating allocations of GRS funds. The relative errors in estimates of postcensal change in per capita income are much larger than those for estimates of total per capita income, for the same reasons cited in our discussion of population estimates for subcounty areas (see sections 1.3b(2) and 2.2c).

There are many problems associated with measuring income trends in small areas. Agriculture or farm income is a more important component of income in areas with smaller populations than in areas with larger populations. In addition, the measurement of entrepreneurial income has conceptual problems not associated with other income figures (see section 5.1e).

Another problem centers on the volatility of agriculture. It is difficult to find a typical year. The Bureau utilizes constraints on estimates of county farm income to damp sharp year-to-year changes, but these constraints will do little to improve accuracy if the volatility is real and not a figment of the data.

Given the inherent difficulty in measuring income changes, especially for areas in which income is largely agricultural, the Panel does not believe that accurate estimates of per capita income can be made for sub-county areas (except, possibly, for very large cities). The Panel is also wary of the county estimates, for which tests of accuracy are not available. Such tests should be performed when the 1980 census results are available.

1.4 RECOMMENDATIONS

RECOMMENDATION 1

The Panel recommends that the Census Bureau continue to make postcensal population estimates for all counties and for all places above a certain size. That certain size, the threshold, should be determined by a systematic evaluation of estimation methods against the 1980 census. The Census Bureau should not make postcensal population estimates for places with population below that threshold.

It is the view of the Panel that the Census Bureau should continue to produce estimates for selected subcounty areas; however, estimates should not continue to be made for subcounty areas that are too small for accurate estimates. Although more evaluation is needed before a precise determination can be made of how small is too small, it is clear that a population of 500 is too small. The average relative error of postcensal estimates of total population for subcounty areas with less than 500 population was 23 percent (based on data for areas that had special censuses taken 1975). For subcounty areas with 1,000–2,499 population the average error in estimates of total population was 10 percent, but this represented an average of 111 percent in the estimate of 1970–1975 change in population of these areas. In our view—with the data available at this time—a population of 5,000 or 10,000 may be a reasonable threshold, but a final determination should await a comparison of postcensal estimates for 1980 with 1980 census counts. In 1975, only 15 percent of the subcounty areas (municipalities and townships) had a population of 5,000 or more, but these areas contained 83 percent of the total population of all subcounty areas. Similarly, more than one-third (36 percent) of all subcounty areas had less than 500 population, but these areas contained less than 2 percent of the total population of subcounty areas.

RECOMMENDATION 2

The Panel recommends that the Census Bureau not make postcensal estimates of per capita money income below the county level. Serious consideration should be given to discontinuing estimates for counties as well, but a decision on this should await comparisons of the postcensal estimates with the 1980 census.

The task of estimating postcensal per capita income is even more formidable than that for population. The limited evaluation data available indicate that the subcounty per capita income estimates are less accurate than the population estimates. No evaluation data were available for the county estimates, but the Panel is suspicious of their accuracy, especially for those counties for which farm income is a significant component of total income. Since the subcounty population and income estimates are used to update general revenue sharing allocations, alternative ways of updating these allocations may need to be considered; some suggestions are given below.

RECOMMENDATION 3

The Panel urges that responsibilities within the Census Bureau be reassigned to bring theoretical and applied statisticians more fully into the estimation program, especially in relation to the development, analysis, and review of estimation procedures. The Bureau should use expertise from within to pursue methodological innovations, and when this expertise is not available, the Bureau should draw upon appropriate talent from outside.

The Panel believes that the postcensal estimation program has not received sufficient attention from theoretical and applied statisticians. The Bureau has successfully applied a few methodological innovations (such as the empirical Bayes estimation methods—see Fay and Herriot (1979)), but there is room for more. Examples of underutilized methods for the postcensal estimation program include variance components models, empirical Bayes estimation, time-series models, and use of diagnostic techniques for model fitting. More research is also needed to develop and extend methodology for evaluation of the estimates. The capabilities for implementing these methods are not sufficiently used in the estimation program at present.

RECOMMENDATION 4

The local estimates of population and per capita income should be given a full statistical evaluation. This should especially include the following: a statement of desired statistical criteria for estimates with justification; investigation of biases for categories of areas; investigation of outliers; investigation of differentially weighted averages of methods; and investigation of errors in estimates of postcensal change in population and per capita income, as well as errors in postcensal estimates of total population and per capita income.

The Panel's evaluation of the accuracy of the estimates has rested largely on comparisons of the estimates with the results of special censuses carried out in the 1970s. For various reasons, the areas for which special censuses were done may not be typical of all areas. More conclusive evaluation of the estimates awaits comparison of the postcensal estimates with the results of the 1980 census. The Panel believes that the use of censuses is the best method of evaluating postcensal estimates and primary emphasis should be placed on the 1980 census results as the standard against which to compare postcensal estimates. In addition, the Bureau should continue to use special censuses for evaluating postcensal estimates (both those conducted for other purposes and those conducted specifically for evaluating postcensal estimates).

A promising method of evaluation uses estimates obtained from large, high-quality sample-survey estimates, such as the Current Population Survey (CPS). These sample-survey estimates need not be highly accurate in themselves if their variances are known. The evaluation of methods would be enhanced if the CPS were redesigned in minor ways to make it more useful for the estimation of the population and per capita income of a sample of counties and large cities. (Such changes are discussed in section 5.2d.)

The systematic evaluation of the estimation methodology that the Panel recommends will be an expensive undertaking, but the Panel feels that sufficient resources should be allocated for this purpose.

RECOMMENDATION 5

> The Panel recommends that the Census Bureau give serious consideration to relaxing its uniformity criterion and, instead, strive to obtain the most accurate estimates.

The Census Bureau currently uses a uniform methodology to produce its estimates. The same methods are used for all counties and for all local areas within any one state, regardless of population size, rate of growth, or unusual age or other composition factor. If data are available for some but not all jurisdictions within a state, the Census Bureau does not incorporate these data (except for special census data) into its own estimates for *any* of the jurisdictions. Relaxation of this uniformity criterion could increase the accuracy of the Bureau's estimates. The Panel recognizes, however, that administrative and political considerations (such as defense against challenges by local areas) were involved in the Bureau's decision to use a uniform methodology. Therefore the Panel feels that the Census Bureau should be the judge of the extent to which it is feasible to relax the uniformity constraint. (Suggested ways of relaxing this constraint are presented in section 5.2b.)

RECOMMENDATION 6

The Census Bureau should prepare a report describing in detail and explaining the rationale for its methodology for postcensal population and per capita income estimation.

The Census Bureau's documentation of its methodology for population estimation is currently scattered and incomplete. Since the methodology is not described in detail elsewhere, the Panel found it necessary to compile such a description, which is found in Appendix A. A model for the kind of report we recommend is *The Current Population Survey: Design and Methodology* (Bureau of the Census, 1978b). The report should include detailed documentation for the methodology, rationale for the methodology—what is being measured, what criteria of accuracy are employed—and evaluations of tests of the methodology.

RECOMMENDATION 7

The Panel recommends that the Census Bureau's Federal-State Co-operative Program for Local Population Estimates (FSCP) be strengthened. In particular, the Census Bureau should seek authorization and funds to provide resources for state activities and for travel and consultation by state personnel with Census Bureau staff.

The demographic programs in the state agencies need to be strong, fiscally and administratively sound, and professionally staffed. Their strengthening would improve and extend the states' capabilities to provide basic data series for local estimates, to evaluate the quality and completeness of their data, and to review estimates generated by the Bureau of the Census.

RECOMMENDATION 8

The Panel recommends that the place of residence question be included in the 1980 IRS individual income tax returns and that funds be provided to process the data obtained by the question.

In January 1979 the Panel urged (in letters to the Director of the Bureau of the Census and the Director of the Office of Revenue Sharing) "that the 1979 IRS income tax returns contain a special question to determine exact place of residence, as was included on 1975 tax returns."[18] This information on the tax returns plays an essential role in the estimation procedures used by the Census Bureau at both the county and subcounty levels (see Appendix A, sections 2.9, 3.9, and 4.1d). The information is important not only for estimating population and income for small areas but also for larger areas (those with populations of 10,000 or more). If that information is not obtained and analyzed, the Bureau's ability to maintain the accuracy of the local estimates for the 1980s will be impaired. It is clearly too late now to collect the needed information on the 1979 returns—with the added advantage that particular year would have had (explained in the letter)—but it is essential that the information be collected in 1980 (or 1981) in order to update the procedure for the allocation of mailing addresses to appropriate places of residence.

Recommendations 1 and 2 have significant implications for general revenue sharing and other programs that use the postcensal estimates of population and income for determining the allocation of funds or other resources. For GRS (as now structured), if postcensal estimates of population or income are not uniformly available for all subcounty jurisdictions in a state, they cannot be used at all. In addition, the reference dates of the data for each variable must be the same for all subcounty units.[19] These rules imply that the distribution of subcounty proportional shares of county area allocations[20] must either (1) remain frozen until the next census or (2) be updated solely on the basis of changes of local adjusted taxes and, to a lesser extent, intergovernmental transfers of revenue. Even if the proportional shares of county area allocations to subcounty areas are fixed, the sizes of the subcounty allocations will change because GRS

---

[18] See Appendix K for copies of the letters.
[19] Possibly, subcounty per capita income estimates pertaining to different time periods could be used for local governments in different counties, provided the time references for all income estimates within each county were the same. Thus special census results for a subcounty area could not be used unless such results were available for all subcounty areas in the state or county.
[20] Determination of county area allocations is an intermediate stage in the application of the general revenue sharing formula. The county area allocation equals (if the effects of floors and ceilings constraining allocations are neglected) the total allocation to the county government and to all eligible local jurisdictions in the county.

allocations are determined in a hierarchical manner: state allocations are determined first, then divided among county areas and the state government, and then the county area allocations are divided among subcounty areas and the county government. Thus the sizes of the subcounty allocations will change because the county-area allocations will be updated (assuming that per capita income estimates for counties are continued).

Under the first option, if a mid-decade census is carried out to provide accurate population and per capita income statistics for small areas, those results could be used to update the subcounty proportional shares of county allocations for GRS every 5 years. If a mid-decade census is not conducted or is of insufficient scope to yield accurate small-area statistics, then those shares would be updated every 10 years. Currently, only one data element in the general revenue sharing formula is not updated between decennial censuses: urbanized population of a state.

It was outside the scope of the Panel's charge to determine which option would result in more accurate allocations for GRS. Because of interrelationships among different data elements, problems can arise if the data refer to different time periods. A hypothetical example can illustrate this point. A subcounty area's allocation, if not constrained by floors or ceilings (see Appendix E), is equal to a fraction of the allocation to the county area. The fraction is proportional to the ratio of the subcounty area's adjusted tax collections to the square of its per capita income, divided by the sum of these ratios for all subcounty units in the county. It is plausible that changes in local tax collections correspond to some extent to changes in per capita income. Suppose percent changes in a local area's tax collections correspond exactly to changes in the square of its per capita income. Then if perfectly accurate data were available, the area's shares of GRS allocations would not change (*ceteris paribus*) even though the per capita income and adjusted taxes data did change. This is clearly the same outcome that would result if no subcounty data were updated. On the other hand, updating adjusted taxes data alone would serve to *increase* allocations to areas with the fastest rising per capita income and to *decrease* allocations to areas with slowly rising or even decreasing per capita incomes. This is surely contrary to the intent of the law. This extreme example is presented not to argue for either option but to illustrate that the question of whether to update some of the data elements and not others is a subtle one and merits careful consideration.

Allocations are targeted on problems. The GRS formula was designed to "put the money where the needs are" (Joint Committee on Internal Revenue Taxation, 1973, p. 2). But while the statistical variables used for allocation form some kind of measure of the problem, they can never yield an exact measure. They are rather, as Bixby (1977) and the Advisory

Commission on Intergovernmental Relations (1977) say, proxies for the variables that the legislators have in mind. In general revenue sharing, for example, inverse income levels (represented by the reciprocal of per capita income) are used to measure need (Joint Committee on Internal Revenue Taxation, 1973, p. 4), but per capita income by itself does not reflect the type of services needed, the distribution of income, or cost-of-living differences, and it is only a limited measure of affluence.

While the Panel recognizes the desirability of using current data for determining allocations, it does not believe that these data can or even should reflect up-to-the-minute changes. The Panel notes that one important allocation—apportionment of seats in the U.S. House of Representatives—requires updates of the data only every 10 years. Furthermore, using frequent updates of the data can even work against the intent of the legislation. As the House Select Committee on Population (U.S. Congress, 1978, p. 7) noted:

Formulas for the distribution of Federal aid typically include a population-size factor. Therefore, if an area loses some of its inhabitants, it is likely to lose funds when it most needs Federal assistance—during the transition to a smaller tax base and changed needs for social services.

The GRS legislation does not require that updated population and income estimates be produced; it only requires that the most recent estimates that the Census Bureau does produce be used for calculating allocations. One way to achieve more extensive updating of the allocations than that recommended by the Panel would be to change the legislation specifying the allocation formulas to take greater advantage of data for large subcounty areas, for which current and relatively accurate updates are available.

For example, in calculating GRS allocations for periods for which postcensal population estimates are desired, all subcounty jurisdictions with populations below a threshold number (e.g., 500, 5,000, or 10,000) would initially be treated as an aggregate. Postcensal population estimates would be prepared by some procedure for each whole aggregate. The allocation to each aggregate would be determined by formula, possibly in the same manner that allocations would be determined for jurisdictions with populations exceeding the threshold. Allocations within each aggregate would be apportioned on the basis of the last decennial census figures for population and income (and possibly other data). This procedure would allow current updates to be used for the larger subcounty areas but not for smaller areas. The example is illustrative and would need further refinement to become operational.[21]

The Panel suggests that, in light of its recommendations that per capita income no longer be updated below the county level and that serious consideration be given to not updating per capita income at the county level, the Bureau of Economic Analysis's county personal income statistics (on a place of residence basis) be considered for general revenue sharing purposes as a possible substitute for county per capita money income. State personal income currently enters into the determination of state allocations, but personal income does not now enter into the substate formula. Since personal income as measured by the Bureau of Economic Analysis has a conceptual basis consistent with the national income and product accounts, it may be a more appropriate proxy than money income in the GRS substate allocation formula. The Panel does note, however, that the Bureau of Economic Analysis's county personal income estimates are untested and may be no more accurate than the Census Bureau's county money income estimates.

Possible reduction of the cost of producing estimates was not the motivation for the Panel's recommendations 1 and 2. Should a mid-decade census provide accurate local area statistics, however, benefit-cost considerations might indicate a further reduction in the scope of the estimation program. In particular, if population estimates are made only for counties and large subcounty areas, it may be possible to reduce the cost of the administrative records method by reducing the extent of the geocoding operation. (Only 9 percent of the subcounty areas had 10,000 or more population in 1975, although those areas contained 74 percent of the total population.)

Sensitivity analyses should be used to explore the effects of alternative ways of producing estimates on the accuracy and timeliness of the estimates and the effects of those qualities in turn on uses of the estimates. Benefit-cost analyses should be done to compare the costs of alternative techniques or conventions with the benefit of their effects on the estimates' uses. Explicit benefit-cost analysis poses difficult problems—such as specification of how much it is worth spending on data to reduce errors in allocation by given amounts—but even if the problems cannot be com-

---

[21] These refinements concern (but are not limited to) how to prepare separate estimates for portions of jurisdictions that straddle two or more counties, how to identify and treat areas that may grow to exceed the threshold during the period of estimation, and how to handle townships and municipalities separately (only 1,823 of the 16,822 townships had a population of 5,000 or more in 1975; see Bureau of the Census, 1978a, Table E).

pletely resolved, their explicit consideration will be informative and useful (see National Research Council, 1976; Spencer, 1980).

Accurate subcounty figures would be provided by a mid-decade census that attempted 100-percent enumeration of the population. The present legislation permits the mid-decade census to be based on a sample. Accurate subcounty estimates can also be obtained from a sample census, but careful consideration should be given to sampling design. Differential sampling rates should be seriously considered, with relatively low rates for the largest areas and high rates for very small areas. Below some threshold it may be more cost effective to obtain population and income data by a complete census rather than by a sample.

The Panel is concerned about the growing amount of legislation that authorizes the distribution of funds or other resources on the basis of postcensal estimates for small areas. Under the present estimation system the errors in those estimates are likely to be large. Alternative approaches that would yield more accurate estimates are either enormously expensive (e.g., annual censuses) or socially repugnant (e.g., a population register). The Panel believes that the Census Bureau should not allow itself to be put in the position of having to defend estimates that are unavoidably subject to large amounts of statistical error. The pressures on the Census Bureau from complaints, challenges, and likely adjudication are detrimental to its efficient operation.

One particularly troublesome use of statistics for allocation purposes is the determination of whether the population of a given area exceeds a threshold number (usually 50,000 or 100,000), in which case the area will become eligible for funds. For cities with populations near these thresholds it is not possible to say with certainty whether the population is above (or below) the threshold. Cities for which the Bureau produces estimates slightly lower than the threshold number are understandably eager to challenge these figures. The Panel notes and endorses recommendation 9 of the Subcommittee on Statistics for the Allocation of Funds (Office of Federal Statistical Policy and Standards, 1978):

That, to minimize the effects of data errors, eligibility cutoffs be such that there is a gradual transition from receiving no allocation to receiving the full formula amount.

REFERENCES

Advisory Commission on Intergovernmental Relations (1977) *Categorical Grants: Their Role and Design.* Washington, D.C.: U.S. Government Printing Office.

Bixby, L. E. (1977) *Statistical Data Requirements in Legislation.* Prepared for the Committee on National Statistics, National Research Council. Washington, D.C.: National Academy of Sciences.

Bureau of the Census (1973a) *Estimates of Coverage of Population by Sex, Race, and Age: Demographic Analysis.* Census of Population and Housing: 1970, Evaluation and Research Program, PHC(E)-4. Washington, D.C.: U.S. Department of Commerce.

Bureau of the Census (1973b) *Federal-State Cooperative Program for Local Population Estimates: Test Results—April 1, 1970.* Current Population Reports, Series P-26, No. 21. Washington, D.C.: U.S. Department of Commerce.

Bureau of the Census (1974) *Estimates of the Population of States With Components of Change, 1970 to 1973.* Current Population Reports, Series P-25, No. 520. Washington, D.C.: U.S. Department of Commerce.

Bureau of the Census (1975a) *Coverage of Population in the 1970 Census and Some Implications for Public Programs.* Current Population Reports, Series P-23, No. 56. Washington, D.C.: U.S. Department of Commerce.

Bureau of the Census (1975b) *1973 Population and 1972 Per Capita Income Estimates for Counties and Incorporated Places in Alabama.* Current Population Reports, Series P-25, No. 546. Washington, D.C.: U.S. Department of Commerce.

Bureau of the Census (1977a) *Developmental Estimates of the Coverage of the Population of States in the 1970 Census.* Current Population Reports, Series P-23, No. 65. Washington, D.C.: U.S. Department of Commerce.

Bureau of the Census (1977b) *Money Income in 1975 of Families and Persons in the United States.* Current Population Reports, Series P-60, No. 105. Washington, D.C.: U.S. Department of Commerce.

Bureau of the Census (1977c) *1973 (Revised) and 1975 Population Estimates and 1972 (Revised) and 1974 Per Capita Income Estimates for Counties, Incorporated Places and Selected Minor Civil Divisions in New Jersey.* Current Population Reports, Series P-25, No. 678. Washington, D.C.: U.S. Department of Commerce.

Bureau of the Census (1978a) *Census of Governments: 1977,* Vol. 1, No. 1, *Governmental Organization.* Washington, D.C.: U.S. Department of Commerce.

Bureau of the Census (1978b) *The Current Population Survey: Design and Methodology.* Technical Paper No. 40. Washington, D.C.: U.S. Department of Commerce.

Bureau of the Census (1978c) *Population Estimates by Race, for States: July 1, 1973 and 1975.* Current Population Reports, Series P-23, No. 67. Washington, D.C.: U.S. Department of Commerce.

Bureau of the Census (1979) *1976 Population Estimates and 1975 and Revised 1974 Per Capita Income Estimates for Counties, Incorporated Places and Selected Minor Civil Divisions in New Jersey.* Current Population Reports, Series P-25, No. 769. Washington, D.C.: U.S. Department of Commerce.

Coleman, E. (1978) Personal income: Some observations on its construction, uses, and adequacy as a subnational income measure. Pp. 29–33 in Bureau of the Census, *Interrelationship Among Estimates, Surveys, and Forecasts Produced by Federal Agencies.* Small-Area Statistics Papers, Series GE-41, No. 4. Washington, D.C.: U.S. Department of Commerce.

Fay, R. E., and Herriot, R. (1979) Estimates of income for small places: An application of James-Stein procedures to census data. *Journal of the American Statistical Association* 74(366):269–277.

Federal Register (1979) Vol. 44, No. 68. April 6, 1979.

Joint Committee on Internal Revenue Taxation (1973) *General Explanation of the State and Local Fiscal Assistance Act and the Federal-State Tax Collection Act of 1972.* Washington, D.C.: U.S. Government Printing Office.

National Commission on Employment and Umemployment Statistics (1979) *Counting the Labor Force*. Washington, D.C.: U.S. Government Printing Office.

National Research Council (1976) *Setting Statistical Priorities*. Report of the Panel on Methodology for Statistical Priorities. Washington, D.C.: National Academy of Sciences.

National Research Council (1978) *Counting the People in 1980: An Appraisal of Census Plans*. Report of the Panel on Decennial Census Plans. Washington, D.C.: National Academy of Sciences.

Office of Federal Statistical Policy and Standards (1978) *Statistical Policy Working Paper 1: Report on Statistics for Allocation of Funds*. Prepared by the Subcommittee on Statistics for Allocation of Funds, Federal Committee on Statistical Methodology. Washington, D.C.: U.S. Department of Commerce.

Ono, M. (1972) Preliminary evaluation of 1969 money income data collected in the 1970 census of population and housing. Pp. 390-396 in *1972 Proceedings of the Social Statistics Section of the American Statistical Association*. Washington, D.C.: American Statistical Association.

Rosenberg, H., and Myers, G. (1977) State demographic centers: Their current status. *The American Statistician* 31(4):141-146.

Spencer, B. (1980) *Benefit-Cost Analysis of Data Used to Allocate Funds*. Lecture Notes in Statistics 3. New York: Springer-Verlag.

U.S. Congress (1972) *On Availability and Adequacy of Certain Statistical Data With Reference to Proposals for Aid to Cities, Counties and States*. Executive Hearing Before the Committee on Ways and Means, House of Representatives, 92nd Congress. Washington, D.C.: U.S. Government Printing Office.

U.S. Congress (1978) *The Use of Population Data in Federal Assistance Programs*. Subcommittee on Census and Population, Committee on Post Office and Civil Service, House of Representatives, 95th Congress, 2nd session. Washington, D.C.: U.S. Government Printing Office.

Zitter, M. (1972) *Report of the Task Force on Use of Administrative Records for Population Estimates*. Washington, D.C.: Bureau of the Census.

Zitter, M., and Word, D. (1973) Use of Administrative Records for Small Area Population Estimates. Paper presented at Annual Meeting of Population Association of America, New Orleans, La.

PART II EVALUATION OF ESTIMATES AND METHODOLOGY

2 Evaluations of Estimates

2.1 METHODS OF EVALUATION USED BY THE CENSUS BUREAU

Evaluations of estimates inform both the producers and the users of the estimates about the strengths and limitations of the estimators (i.e., estimation procedures). Evaluations help users determine how well the estimates meet their needs. If the estimates are highly accurate (or inaccurate), users may use them (or decline to use them) in making important decisions. If the users are paying for the estimates, they may find less costly but less accurate estimates to be acceptable, or if evaluations show the estimates to be inaccurate, users may decide to seek better estimates even though the expense may be greater. Evaluations aid producers in their efforts to correct weaknesses in existing estimators, design new estimators, average or otherwise combine different estimators, and accept or reject estimators.

The Census Bureau relies primarily on four methods to conduct its evaluations of population estimates: comparison of estimates with decennial census results, comparison of estimates with special census results, comparison among alternative estimates, and use of demographic and statistical logic (Bureau of the Census, 1974, p. 15).[1] We note that decennial censuses are too infrequent for evaluation purposes and that special censuses may give a distorted overall picture because the areas receiving

---

[1] Demographic and statistical logic focuses on whether the assumptions underlying a procedure conform to a logical model of how demographic changes occur.

them are self-selected. Comparisons among alternative estimators are better suited to qualitative than quantitative analysis because one usually
does not know enough about the respective statistical properties (e.g.,
degree or kind of bias) of the estimators to draw precise inferences. If
more were known about their statistical properties, one could make timely
estimates of the levels and directions of error of the estimates.

2.1a COMPARISON WITH DECENNIAL CENSUSES

To use decennial census results to evaluate postcensal estimates, the
Bureau prepares postcensal estimates for a date for which decennial census results are or will be available. The postcensal estimates and census
results are then compared; discrepancies are attributed to errors in the
postcensal estimates, although some of the discrepancies—or lack thereof—may arise from errors in the census figures (see Appendix I for further
discussion). A drawback with using decennial census figures is their infrequent appearance. The performance of estimators can only be evaluated
and compared for 10-year intervals. Little is known about the behavior of
estimators as the time interval for which change is being estimated increases. Even when the average error is known for estimates of 10-year
change, there is still uncertainty about the average error when the time interval is only a few years. Is the variance of a 5-year estimate one-half that
of a 10-year estimate? One-quarter? Seven-tenths? When mid-decade
census results become available, this problem will ease somewhat.

Two decennial censuses may be used as benchmarks for evaluation purposes, one at the beginning and one at the end of a 10-year period. The
former is used prospectively, for making inferences about the accuracy of
estimates in the current postcensal period. For example, to test the ratio-
correlation method and component method II for use in the 1970s, the
Bureau calculated 1970 postcensal estimates based on the census and
symptomatic data for 1960 to 1970 and then compared those estimates
with the results of the 1970 census.

There are two problems in using such comparisons for making inferences about the actual accuracy of the methods for estimating population after 1970. First, demographic processes continually evolve, so that
methods that performed well in the 1960s may perform poorly in the
1970s. In particular, assumptions valid in one decade may not be valid in
the next, relationships among variables may shift over time, and the quality of the available data may also change.

Second, the 1970 census data that are used for gauging the accuracy of
the methods are also used for forming and selecting modifications to the
methods for use in the 1970s. (A good example of this latter use is the

modification of the ratio-correlation method to allow for trends in coverage ratios; see Bureau of the Census (1974) and section 2.5 of Appendix A for details.) Using the same data to evaluate and modify a method and then to evaluate the modification can lead to overestimation of the modified method's accuracy. Methods of cross-validation (Mosteller and Tukey, 1977) can avoid this problem and should be explored. For example, decennial census data for half of the areas could be used to modify the methods, and the data for the other half could be used to assess the accuracy of the modified methods.

2.1b COMPARISON WITH SPECIAL CENSUSES

Use of special censuses avoids some of the problems involved with using decennial censuses, and a more continuous monitoring is possible. Special censuses are censuses conducted at the request and expense of municipalities or counties within a state; they are not part of a national effort. Special censuses can occur throughout the decade, and so the performance of the estimators can be observed continuously over time.

A problem with drawing inferences about overall error rates from comparisons with special censuses arises from selection biases. For example, the places receiving special censuses tend to have higher-than-average growth rates, and it is known that, other things being equal, postcensal population estimation methods are less accurate for fast-growing places than for moderately growing places.[2] Also, special censuses do not generally yield estimates of per capita income. Exceptions to selection bias and absence of questions on income occurred in the sample of 86 special censuses conducted in 1973 by the Census Bureau to permit evaluation of the estimation methodologies for population and per capita income. The Bureau paid for these censuses and selected the areas so as to constitute a probability sample of local areas with less than 20,000 population. Special censuses conducted specifically for evaluation purposes provide the strongest evaluation and should be adopted when resources permit. However, it is probably prohibitively expensive to take a sufficiently large number of special censuses to yield conclusive results.

It must be recognized that decennial or special censuses are themselves subject to error, a fact that should be taken into consideration in evaluations. Thus the difference between a census enumeration and a postcensal estimate is not in general just the error in the postcensal estimate. (Appen-

---

[2] Regression analysis or other techniques of data analysis should be useful for disentangling the effects of these biases, but modern techniques of data analysis have so far seen little application to the special census comparisons.

dix I suggests one explicit approach to take into account undercoverage in census enumerations.) Error in adjusted census counts used as the base for evaluation also arises because such counts are usually linearly interpolated or extrapolated over time to attain agreement with the reference date of the postcensal estimates.

2.1c COMPARISON AMONG ALTERNATIVE ESTIMATES

An alternative to using decennial census or special census figures is to compare estimates produced by various methods or to use other data series. For instance, several estimates (obtained from the AR, RC, CM II, and possibly other methods) of postcensal county populations can be compared, and the dispersion of the estimates studied. Unfortunately, this information is of limited utility because the true population size is unknown: the alternative estimates may all lie above the true value, they may all lie below it, or they may straddle it. An increase in the extent of dispersion over time is often taken as being indicative of the deterioration of the accuracy of one or more of the estimators over time, but the indication is only a weak one.

2.1d DEMOGRAPHIC AND STATISTICAL LOGIC

A fourth method of evaluation used by the Census Bureau is consideration of the demographic and statistical logic of the assumptions underlying the estimation methods, along with judgment. For example, some have argued that the AR method systematically underestimates the populations of large central cities (Mann, 1978). This argument is based on logic rather than on statistical evaluation of the estimates produced by the method (see section 5.1b(1) for further discussion). Judgment focuses on the plausibility of the output of the methods. For example, Census Bureau staff use judgment to decide when to stop incorporating information from past special censuses into current population estimates provided by the CM II or RC method (see Appendix A, section 3.11). Appendix C further analyzes the role of judgment in making postcensal estimates.

2.2 PANEL EVALUATION OF POPULATION ESTIMATES

This section analyzes empirical evidence about the accuracy of postcensal population estimates. Our discussion focuses largely on comparisons of the estimates with results of special censuses, where available.[3] As discussed above, deviations of the estimates from adjusted special census counts are not precise reflections of errors in the estimates because the special census counts contain error from misenumeration and from adjustment (interpolation or extrapolation to the date of the population estimate). Furthermore, one must be guarded in drawing inferences from these comparisons because the areas receiving special censuses are usually self-selected; they are not a random sample of all areas for which estimates are made.

In comparing estimates with the special census counts, we focus on three of the criteria of accuracy discussed in section 1.1d:

Criterion 2—Low average *relative* error. Relative error is measured by the difference between the population estimate and the special census count, expressed as a percentage of the count (hereafter referred to simply as "percent difference"). Average relative error refers to the arithmetic mean of percent differences *disregarding sign*.

Criterion 3—Few extreme *relative* errors. Extreme error is measured by the proportion of percent differences exceeding a specified value, disregarding sign, often 10 or 15 percent.

Criterion 4—Bias. Bias is measured in terms of the numbers of areas whose estimates exceed the special census counts (positive differences) and fall below the special census counts (negative differences).

These are the same criteria used by the Census Bureau in its evaluation of state and county population estimates against the 1970 census results (Bureau of the Census, 1973b).

Criterion 4 provides information about bias in the population estimates for a group of areas—an excessive number of positive differences suggests an upward bias in the estimation methodology for the group of areas. It is important to remember, however, that the subcounty estimates are controlled to those for larger geographic areas, so if the estimate for a county is too high, the subcounty estimates for that county may appear to be biased upward even if they are unbiased estimates of the proportions of county population living in the subcounty areas.

2.2a STATE POPULATION ESTIMATES

Tests reported by the Bureau of the Census (1974) show that 1970 estimates for states (derived from the 1960 census and symptomatic data for

---

[3]The special censuses that were used include those conducted by the Census Bureau (summarized in *Current Population Reports,* Series P-28) and also those conducted by state or local agencies and accepted by the Census Bureau.

1960 to 1970, then compared with 1970 census counts) had very high levels of accuracy for the three methods used by the Census Bureau. State estimates produced by averaging the results of component method II and the ratio-correlation method deviated from the 1970 census counts by an average of only 1.2 percent. Similar estimates for 1970 could not be prepared by the AR method because the necessary symptomatic data were not available for 1960 to 1970, but comparison of 1975 AR method estimates for states with comparable estimates by the other two methods showed that the average percent difference between the AR estimate and the average of CM II and RC estimates was only 1 percent, with the vast majority of states having differences of less than 2 percent (Bureau of the Census, 1980).

2.2b COUNTY POPULATION ESTIMATES

Estimates of county population, in terms of average error, are quite accurate. The average percent difference between postcensal estimates and special census counts was 3.9 percent for 133 counties in which special censuses were taken during the 3-year period of 1974 through 1976 (Table 2.1). The accuracy of the estimates varied with the 1970 population size of the county and with the percent change in population size between 1970 and the special census date: large counties were estimated more accurately than small counties; slowly growing counties were estimated more accurately than declining or rapidly growing counties. For example, the average percent difference between postcensal estimates and special census counts decreased from 7.1 for counties with 1970 population of less than 1,000 to 1.4 for counties with 1970 population of 100,000 or more. The average percent differences for large counties (25,000 or more population) were small, 3 percent or less, regardless of rate of growth or decline.

When counties were classified by rate of change in population, the most accurate estimates were for those counties with moderate growth: the average percent difference was 2.0 for counties that grew by 5 to 10 percent and 2.7 for those that grew by 10 to 15 percent. Counties that had grown by 25 percent or more had substantially larger errors (an average percent difference of 6.8), as did counties that had declined in population by 5 percent or more (an average percent difference of 5.1). The largest errors occurred among counties that were both small and experiencing rapid growth (Table 2.2). The average percent difference reached a high of 11.2 percent among counties that had less than 5,000 population in 1970 and had grown by 15 percent or more since 1970.

Because the county estimates are, on the whole, quite accurate, it is not

TABLE 2.1 Percent Difference Between Postcensal Estimates of Population and Special Census Counts, by 1970 Population and Percent Change in Population Since 1970: 133 Counties With Special Censuses Taken Between January 1, 1974, and December 31, 1976

1970 Population and Percent Change in Population Since 1970	Number of Counties	Average Percent Difference[a]	Percent of Counties With Difference of	
			5.0 Percent or More[a]	10.0 Percent or More[a]
All counties	133	3.9	23	8
By 1970 Population				
Less than 1,000	24	7.1	50	25
1,000 to 4,999	23	5.2	30	17
5,000 to 9,999	12	3.6	33	0
10,000 to 24,999	20	3.6	25	5
25,000 to 49,999	8	3.5	25	0
50,000 to 99,999	14	2.5	7	0
100,000 or more	32	1.4	0	0
By Percent Change Since 1970				
−5.0 percent or more	11	5.1	27	18
−0.0 to −4.9 percent	16	4.0	31	6
+0.0 to +4.9 percent	25	3.3	20	8
+5.0 to +9.9 percent	19	2.0	5	0
+10.0 to +14.9 percent	23	2.7	13	0
+15.0 to +24.9 percent	17	3.6	24	6
+25.0 percent or more	22	6.8	45	23

[a] Percent difference for each county equals postcensal estimate (as of July 1) *minus* adjusted special census count (interpolated or extrapolated to July 1 of year special census was taken), expressed as percent of adjusted census count. Average percent difference calculated as arithmetic mean of percent differences *disregarding sign*. Counties with differences of 5.0 (or 10.0) percent or more were tallied *disregarding sign* of differences.

SOURCE: Unpublished data from the Bureau of the Census provided by Frederick Cavanaugh.

surprising that extreme errors are relatively uncommon except among small counties or counties undergoing rapid growth or decline. Of all counties with less than 1,000 population, 25 percent had errors of 10 percent or more, as did 17 percent of the counties with 1,000 or 4,999 population, 23 percent of the counties that had grown by 25 percent or more since 1970, and 18 percent of the counties that had declined in population by 5 percent or more (Table 2.1). Extreme errors seldom occurred among counties with larger populations or slower rates of change.

TABLE 2.2   Percent Difference Between Postcensal Estimates of Population and Special Census Counts, by 1970 Population Cross-Classified by Percent Change in Population Since 1970: 133 Counties With Special Censuses Taken Between January 1, 1974, and December 31, 1976

1970 Population	Total	Percent Change in Population Since 1970				
		−5.0 or More	−0.0 to −4.9	+0.0 to +4.9	+5.0 to +14.9	+15.0 or More
All Counties						
Average percent difference[a]	3.9	5.1	4.0	3.3	2.4	5.4
Number of counties	133	11	16	25	42	39
Total with positive differences	65	8	10	17	23	7
Less Than 5,000						
Average percent difference[a]	6.2	5.1	4.3	5.5	3.1	11.2
Number of counties	47	11	9	9	7	11
Total with positive differences	25	8	5	6	4	2
5,000 to 24,999						
Average percent difference[a]	3.6	—	4.6	3.4	2.9	4.4
Number of counties	32	0	4	4	14	10
Total with positive differences	16	—	3	3	7	3
25,000 to 99,999						
Average percent difference[a]	2.9	—	—	2.5	3.2	3.1
Number of counties	22	0	0	5	9	8
Total with positive differences	9	—	—	3	4	2
100,000 or More						
Average percent difference[a]	1.4	—	2.5	1.4	0.8	1.8
Number of counties	32	0	3	7	12	10
Total with positive differences	15	—	2	5	8	0

[a] Percent difference for each county equals postcensal estimate (as of July 1) *minus* adjusted special census count (interpolated or extrapolated to July 1 of year special census was taken), expressed as percent of adjusted census count. Average percent difference calculated as arithmetic mean of percent differences *disregarding sign*.

SOURCE: Unpublished data from the Bureau of the Census provided by Frederick Cavanaugh.

There also is evidence of bias in the estimation methods: they tend to overestimate the population of declining counties and to underestimate the population of rapidly growing counties. For example, the estimates were too high for 8 of 11 counties that had declined in population by 5 percent or more since 1970, while estimates were too low for 32 of 39 counties that had grown by 15 percent or more (Table 2.2). That is, the estimation methods tend to underestimate the change in population, for both declining and increasing populations.

The patterns noted above also hold, in general, for each of the three individual methods that are averaged to obtain the postcensal estimates (Table 2.3). The error is, on the average, lower for the postcensal estimate than for the individual methods. It should be noted, however, that the difference between the average error of the postcensal estimates and the AR estimates is very small and that the average error for the postcensal estimates is not consistently lower in all the population-size and rate-of-growth subgroups. This result can occur because the different methods that are averaged are not equally accurate. Of the three individual methods the average percent difference is lowest for the AR method (4.0), next lowest for the RC method (4.9), and highest for CM II (6.4). The proportion of percent differences (disregarding sign) that are 10 percent or more is smallest (8 percent) for the AR method and the postcensal estimate, next smallest (13 percent) for the RC method, and largest (21 percent) for CM II (calculated from Tables 2.4 and 2.5).

The accuracy of the county estimates is also affected by the age structure of the population. Table 2.6 shows that the estimates are less accurate for counties whose populations had age distributions dissimilar to the age distribution for the nation as a whole. The index of dissimilarity (Δ), the measure of dissimilar age structure used in Table 2.6, is defined as

$$\Delta_j = \frac{1}{2} \sum_i |P_{ij} - p_i|,$$

where P_{ij} is the percentage of county j population in age category i and p_i is the percentage of the total national population in age category i. (The age categories used are 0–17 years, 18–64 years, and 65 and over). With one exception the average percent differences are larger for counties with dissimilar age structure, that is, with Δ of 5 percent or more. The exception is the RC estimates for small counties (less than 5,000 population), where the average percent difference between the estimates and special census counts was lower for counties with dissimilar age structure. But even for this case the proportion of differences (disregarding sign) that exceeded 5 percent was higher for the counties with dissimilar age structure.

TABLE 2.3   Average Percent Difference Between Population Estimates and Special Census Counts for Four Different Methods of Estimation, by 1970 Population and Percent Change in Population Since 1970: 133 Counties With Special Censuses Taken Between January 1, 1974, and December 31, 1976

1970 Population and Percent Change in Population Since 1970	Number of Counties	Average Percent Difference[a]			
		Post-censal Estimate[b]	Component Method II	Ratio-Correlation	Administrative Records Method
All Counties	133	3.9	6.4	4.9	4.0
By 1970 Population					
Less than 1,000	24	7.1	13.6	8.8	6.8
1,000 to 4,999	23	5.2	8.8	7.2	5.3
5,000 to 9,999	12	3.6	5.4	5.0	5.3
10,000 to 24,999	20	3.6	5.6	3.7	3.3
25,000 to 49,999	8	3.5	4.1	5.5	2.8
50,000 to 99,999	14	2.5	3.6	3.4	3.2
100,000 or more	32	1.4	2.0	1.7	1.7
By Percent Change Since 1970					
−5.0 percent or more	11	5.1	11.8	5.7	7.5
−0.0 to −4.9 percent	16	4.0	6.2	6.0	4.0
+0.0 to +4.9 percent	25	3.3	7.0	4.3	2.9
+5.0 to +9.9 percent	19	2.0	3.8	3.2	2.1
+10.0 to +14.9 percent	23	2.7	4.1	3.0	3.4
+15.0 to +24.9 percent	17	3.6	6.2	3.8	2.6
+25.0 percent or more	22	6.8	8.0	8.8	7.1

[a] Percent difference for each county equals postcensal estimate (as of July 1) *minus* adjusted special census count (interpolated or extrapolated to July 1 of year special census was taken), expressed as percent of adjusted census count. Average percent difference calculated as arithmetic mean of percent differences *disregarding sign*.
[b] Calculated as average of estimates obtained by the three methods (in some states, also includes a fourth estimate prepared by the states).

SOURCE: Unpublished data from the Bureau of the Census provided by Frederick Cavanaugh.

Simple stochastic models for error in the estimates lead one to believe that the error in the estimates will increase as the length of time since the last decennial census increases. However, the hypothesis of increasing error over time is difficult to test with the available data because the areas receiving special censuses are self-selected, so that differences in estimated error levels over time may be largely attributable to changes in the set of areas receiving special censuses. One way to evaluate the hypothesis in intercensal years would be to use the regression method of Ericksen (1974) to estimate the level of error. Since this method does not rely on special censuses, it can be used to analyze changes in the level of accuracy. Another way to study the behavior of error over time is to analyze the behavior of alternative estimators, relative to each other, over time, but this approach may be misleading because the estimates from different methods may remain close to each other but be far from the actual population value (see Voss, 1978). After the 1980 census has been taken, other evaluations of the hypothesis will be possible (see below).

2.2c SUBCOUNTY ESTIMATES

Estimates of the population of subcounty areas in 1975 were quite accurate for areas with large populations but were increasingly inaccurate as population size decreased. For example, the average percent difference between 1975 population estimates and comparable 1975 special census counts was only 2.6 to 2.7 percent for areas with 25,000 or more population in 1970 but increased to more than 25 percent for areas that had less than 250 population in 1970 (Table 2.7).

The accuracy of the estimates also varied greatly by the rate at which the population was changing from 1970 to 1975. Areas with relatively stable populations—less than 5 percent growth or decline—had an average percent difference of 6 percent, as compared with areas that grew by at least 50 percent or that declined by at least 10 percent, which had an average percent difference of more than 20 percent.

The strong patterns exhibited in Table 2.7—increasing error with decreasing size of population and increasing error with increasing rate of change in population size—persist when measures of accuracy are cross-classified by both variables simultaneously, as shown in Table 2.8. Estimates for areas that were both small and subject to rapid growth or decline were most inaccurate. For example, the average error was 43 percent for areas that had less than 500 population in 1970 *and* whose population had declined by 10 percent or more between 1970 and 1975. The average error for areas that grew by 50 percent or more from 1970 to 1975 was high in all population-size groups except the largest: the average percent difference decreased from a high of 27 percent for areas with less than 500 population to 19 percent for areas with 10,000 to 24,999 population and declined sharply to 7 percent for areas with 25,000 or more population. Similarly, very small areas (those with less than 500 population) had large errors regardless of the rate of change in population size:

TABLE 2.4 Distribution of Counties by Percent Difference Between Population Estimates and Special Census Counts, by 1970 Population: 133 Counties With Special Censuses Taken Between January 1, 1974, and December 31, 1976

Size of Percent Difference[a]	Total	1970 Population						
		Less Than 1,000	1,000–4,999	5,000–9,999	10,000–24,999	25,000–49,999	50,000–99,999	100,000 or More
Postcensal Estimates[b]								
TOTAL	133	24	23	12	20	8	14	32
−10.0 to −24.9 percent	6	4	2	0	0	0	0	0
−5.0 to −9.9 percent	10	2	1	3	3	0	1	0
−0.0 to −4.9 percent	52	8	5	2	8	5	7	17
+0.0 to +4.9 percent	50	4	11	6	7	1	6	15
+5.0 to +9.9 percent	10	4	2	1	1	2	0	0
+10.0 to +24.9 percent	5	2	2	0	1	0	0	0
Ratio-Correlation Method								
TOTAL	133	24	23	12	20	8	14	32
−25.0 percent or more	1	0	1	0	0	0	0	0
−10.0 to −24.9 percent	8	6	1	0	0	0	1	0
−5.0 to −9.9 percent	15	4	4	2	3	0	1	1
−0.0 to −4.9 percent	46	6	3	4	9	4	5	15
+0.0 to +4.9 percent	40	4	6	2	5	1	6	16
+5.0 to +9.9 percent	15	1	6	4	2	1	1	0
+10.0 to +24.9 percent	8	3	2	0	1	2	0	0

Component Method II

	133	24	23	12	20	8	14	32
TOTAL	133	24	23	12	20	8	14	32
−10.0 to −24.9 percent	15	7	5	2	1	0	0	0
−5.0 to −9.9 percent	16	0	3	1	6	1	2	3
−0.0 to −4.9 percent	47	4	6	4	4	5	7	17
+0.0 to +4.9 percent	31	4	3	3	6	0	3	12
+5.0 to +9.9 percent	11	3	2	1	1	2	2	0
+10.0 to +24.9 percent	8	2	3	1	2	0	0	0
+25.0 percent or more	5	4	1	0	0	0	0	0

Administrative Records

	133	24	23	12	20	8	14	32
TOTAL	133	24	23	12	20	8	14	32
−25.0 percent or more	1	1	0	0	0	0	0	0
−10.0 to −24.9 percent	7	4	2	1	0	0	0	0
−5.0 to −9.9 percent	8	2	0	2	3	0	0	1
−0.0 to −4.9 percent	48	5	5	3	11	5	5	14
+0.0 to +4.9 percent	50	8	9	4	4	2	6	17
+5.0 to +9.9 percent	16	2	6	2	2	1	3	0
+10.0 to +24.9 percent	3	2	1	0	0	0	0	0

[a]Percent difference for each county equals postcensal estimate (as of July 1) *minus* adjusted special census count (interpolated or extrapolated to July 1 of year special census was taken), expressed as percent of adjusted census count.

[b]Calculated as average of estimates obtained by the three methods (in some states, also includes a fourth estimate prepared by a state agency).

SOURCE: Unpublished data from the Bureau of the Census provided by Frederick Cavanaugh.

TABLE 2.5 Distribution of Counties by Percent Difference Between Population Estimates and Special Census Counts, by Percent Change in Population Since 1970: 133 Counties With Special Censuses Taken Between January 1, 1974, and December 31, 1976

Size of Percent Difference[a]	Total	Percent Change in Population, 1970 to Special Census						
		Decrease		Increase				
		−5.0 Percent or More	−0.0 to −4.9 Percent	+0.0 to +4.9 Percent	+5.0 to +9.9 Percent	+10.0 to +14.9 Percent	+15.0 to +24.9 Percent	+25.0 Percent or More
Postcensal Estimates[b]								
TOTAL	133	11	16	25	19	23	17	22
−10.0 to −24.9 percent	6	0	0	1	0	0	0	5
−5.0 to −9.9 percent	10	0	1	0	0	1	3	5
−0.0 to −4.9 percent	52	3	5	7	7	11	11	8
+0.0 to +4.9 percent	50	5	6	13	11	9	2	4
+5.0 to +9.9 percent	10	1	3	3	1	2	0	0
+10.0 to +24.9 percent	5	2	1	1	0	0	1	0
Ratio-Correlation Method								
TOTAL	133	11	16	25	19	23	17	22
−25.0 percent or more	1	0	0	0	0	0	0	1
−10.0 to −24.9 percent	8	0	1	2	0	0	1	4
−5.0 to −9.9 percent	15	2	1	1	4	2	0	5
−0.0 to −4.9 percent	46	4	4	6	5	8	11	8

+0.0 to +4.9 percent	40	2	6	10	8	10	4	0
+5.0 to +9.9 percent	15	1	2	5	2	2	0	3
+10.0 to +24.9 percent	8	2	2	1	0	1	1	1
Component Method II								
TOTAL	133	11	16	25	19	23	17	22
−10.0 to −24.9 percent	15	2	2	1	0	1	1	8
−5.0 to −9.9 percent	16	0	2	2	2	2	3	5
−0.0 to −4.9 percent	47	2	4	10	5	11	9	6
+0.0 to +4.9 percent	31	2	4	5	10	5	3	2
+5.0 to +9.9 percent	11	1	3	3	1	3	0	0
+10.0 to +24.9 percent	8	2	1	3	0	1	0	1
+25.0 percent or more	5	2	0	1	1	0	1	0
Administrative Records								
TOTAL	133	11	16	25	19	23	17	22
−25.0 percent or more	1	0	0	0	0	0	0	1
−10.0 to −24.9 percent	7	0	0	1	0	0	1	5
−5.0 to −9.9 percent	8	1	1	0	0	3	1	2
−0.0 to −4.9 percent	48	1	3	9	8	11	7	9
+0.0 to +4.9 percent	50	2	8	14	9	5	7	5
+5.0 to +9.9 percent	16	4	4	1	2	4	1	0
+10.0 to +24.9 percent	3	3	0	0	0	0	0	0

[a] Percent difference for each county equals postcensal estimate (as of July 1) *minus* adjusted special census count (interpolated or extrapolated to July 1 of year special census was taken), expressed as percent of adjusted census count.

[b] Calculated as average of estimates obtained by the three methods (in some states, also includes a fourth estimate prepared by the state).

SOURCE: Unpublished data from the Bureau of the Census provided by Frederick Cavanaugh.

TABLE 2.6 Percent Difference Between Population Estimates and Special Census Counts for Four Different Methods of Estimation, by 1970 Population and 1970 Age Structure: 133 Counties With Special Censuses Taken Between January 1, 1974, and December 31, 1976

| Type of Estimate and 1970 Population | Index of Dissimilarity of Age Structure[a] | | | | | | Number of Counties | |
| | Average Percent Difference[b] | | | Percent of Counties with Differences of 5 Percent or More[c] | | | | |
	Under 5.0 Percent	5.0 Percent or More	Total	Under 5.0 Percent	5.0 Percent or More	Total	Under 5.0 Percent	5.0 Percent or More
Postcensal Estimate[d]								
TOTAL	3.2	4.9	3.9	16	33	23	79	54
Less than 5,000	5.8	6.6	6.2	35	48	40	26	21
5,000 to 99,999	2.6	4.0	3.3	15	30	22	27	27
100,000 or more	1.1	2.5	1.4	0	0	0	26	6
Administrative Records								
TOTAL	3.6	4.7	4.0	20	35	26	79	54
Less than 5,000	6.0	6.1	6.1	38	48	43	26	21
5,000 to 99,999	3.3	3.9	3.6	22	30	26	27	27
100,000 or more	1.3	3.4	1.7	0	17	3	26	6

Component Method II

TOTAL	5.6	7.6	6.4	34	52	41	79	54
Less than 5,000	10.5	12.1	11.2	58	72	64	26	21
5,000 to 99,999	4.7	4.9	4.8	41	41	41	27	27
100,000 or more	1.7	3.6	2.0	4	33	9	26	6
Ratio-Correlation								
TOTAL	4.1	6.1	4.9	24	52	35	79	54
Less than 5,000	8.1	7.9	8.0	54	67	60	26	21
5,000 to 99,999	3.0	5.4	4.2	19	48	33	27	27
100,000 or more	1.4	2.6	1.7	0	17	3	26	6

[a] The index of dissimilarity of age structure is equal to one-half the sum of the absolute values of differences between corresponding percents of two age distributions, in this case the age distribution of the county and the age distribution of the nation in 1970. Three age categories were used in the calculation: under age 18, ages 18–64, age 65 and older.

[b] Percent difference for each county equals postcensal estimate (as of July 1) *minus* adjusted special census count (interpolated or extrapolated to July 1 of year special census was taken), expressed as percent of adjusted census count. Average percent difference calculated as arithmetic mean of percent difference *disregarding sign*.

[c] Counties with differences of 5.0 percent or more were tallied *disregarding sign* of differences.

[d] Calculated as average of estimates obtained by the three methods (in some states, also includes a fourth estimate prepared by the states).

SOURCE: Age data from Bureau of the Census (1973a). Other data from the Bureau of the Census provided by David Word.

TABLE 2.7   Percent Difference Between 1975 Estimates of Population and Special Census Counts, by 1970 Population and Percent Change in Population, 1970-1975: 799 Subcounty Areas With Special Censuses Taken During 1975

| | | | | Percent of Areas With | | |
| | | | | Percent Differences (Positive or Negative) of | | |
1970 Population and Percent Change in Population, 1970-1975	Average Percent Difference[a]	Number of Areas	Positive Differences[b]	10 Percent or More	15 Percent or More	25 Percent or More
All Subcounty Areas	11.7	799	44	34	24	12
By 1970 Population						
Less than 50	26.0	33				
50 to 249	27.1	123	46	66	52	30
250 to 499	13.5	67				
500 to 999	9.8	77	43	32	17	6
1,000 to 2,499	10.3	118	36	33	21	8
2,500 to 4,999	7.2	111	39	21	15	5
5,000 to 9,999	8.2	88	45	26	17	5
10,000 to 24,999	5.2	94	45	10	7	3
25,000 to 49,999	2.6	50	54	2	2	0
50,000 to 99,999	2.6	27	50	5	0	0
100,000 or more	2.7	11				
By Percent Change, 1970-1975						
−25.0 percent or more	83.8	15	84	77	65	47
−10.0 to −24.9 percent	22.7	42				
−5.0 to −9.9 percent	9.5	38	72	23	15	6
−0.0 to −4.9 percent	6.2	77				
+0.0 to +4.9 percent	6.4	114	60	18	13	5
+5.0 to +9.9 percent	6.9	104	41	23	12	3
+10.0 to +24.9 percent	7.5	228	34	25	14	5
+25.0 to +49.9 percent	12.0	105	22	44	30	7
+50.0 percent or more	24.1	76	9	68	66	43

[a] Percent difference for each area equals postcensal estimate as of July 1, 1975, *minus* adjusted special census count (interpolated or extrapolated to July 1, 1975), expressed as percent of adjusted census count. Average percent difference calculated as arithmetic mean of percent differences *disregarding sign.*

[b] Percent based on total number of areas with positive *or* negative difference (that is, total excluding areas for which the estimate was exactly equal to the adjusted census count); 11 of the 799 postcensal estimates were exactly equal to the adjusted census counts.

SOURCE: Unpublished data from the Bureau of the Census provided by Frederick Cavanaugh.

the average percent difference was 13 to 15 percent for areas with moderate growth or decline and 27 and 43 percent for areas of fast growth or decline. Only among areas with 25,000 or more population in 1970 were the estimates relatively accurate regardless of rate of change in population: the average percent difference for these areas was 2.4 percent among those that changed (growth or decline) by less than 10 percent and 6.6 percent for areas that grew by 50 percent or more.

For all 799 subcounty areas (municipalities and townships) in which special censuses were taken during 1975 and compared with 1975 population estimates, the overall average difference was 11.7 percent (Table 2.9). This overall average, however, reflects the composition of the largely self-selected group of subcounty areas in which special censuses were taken and may be different from the average for the more than 35,000 municipalities and townships eligible for general revenue sharing (GRS).[4] For example, only 38 percent of the 799 special census areas had less than 1,000 population in 1970 as compared with 54 percent of the full set of sub-

---

[4] The 799 subcounty areas for which data are reported in Tables 2.7–2.9 include 426 in which special censuses were taken by the Census Bureau in 1975 and 373 in which special censuses were taken by state or local agencies and accepted by the Bureau. The computer printout list from which the tables were compiled was provided by the Census Bureau, but we did considerable editing prior to our tabulations.

The computer printout included all special censuses that were adjusted to July 1, 1975 (by interpolation or extrapolation) and compared with 1975 population estimates; some of these censuses were taken in years other than 1975. The printout also had separate listings for "balances" of townships that included a municipality and for separate pieces of municipalities that straddled township or county boundaries. In all, there were 1,544 comparisons with 1975 estimates on the printout, but 345 of them were based on a single special census of the entire state of Massachusetts, which was taken by the state government on March 1, 1975. Rather than have our comparisons dominated by one special census (of unknown quality) covering every subcounty area in one state, we decided to exclude the comparisons for Massachusetts. In our editing we also dropped 272 comparisons that were based on extrapolated counts of special censuses taken in 1974, 56 comparisons for "County Balances," 30 comparisons for areas in which special censuses were taken in another year or in both 1974 and another year, and one comparison for which we could not identify the place code. We also combined separate pieces of municipalities that straddled township boundaries (47 pieces were combined into 18 municipalities) or county boundaries (23 pieces combined into 11 municipalities), and substituted 14 township totals for 14 "Township Balances" that were listed separately on the computer printout.

The objectives of the editing process were to obtain a set of comparisons of 1975 population estimates with adjusted special census counts for a set of subcounty areas defined on an equivalent basis to general revenue sharing governmental jurisdictions (whole jurisdictions) and to limit the adjustment period (for interpolation or extrapolation) to less than 6 months. The second objective led us to exclude from our tables comparisons of 1975 estimate with adjusted counts (as of July 1, 1975) interpolated or extrapolated from special censuses taken in any year other than 1975.

TABLE 2.8 Percent Difference Between 1975 Estimates of Population and Special Census Counts, by 1970 Population Cross-Classified by Percent Change in Population, 1970-1975: 799 Subcounty Areas With Special Censuses Taken During 1975

1970 Population of Subcounty Area	Total	Percent Change in Population, 1970-1975					
		−10.0 or More	−0.0 to −9.9	+0.0 to +9.9	+10.0 to +24.9	+25.0 to +49.9	+50.0 or More
All Subcounty Areas							
Average percent difference	11.7	38.8	7.3	6.6	7.5	12.0	24.1
Number of areas—TOTAL	799	57	115	218	228	105	76
With positive differences	344	46	82	109	77	23	7
With (±) differences of							
10.0 percent or more	270	44	27	45	56	46	52
15.0 percent or more	195	37	17	27	33	31	50
25.0 percent or more	95	27	7	9	12	7	33
Areas with Less Than 500 Population							
Average percent difference	22.9	42.9	13.7	15.1	13.5	20.0	27.4
Number of areas—TOTAL	223	50	25	42	53	25	28
With positive differences	98	41	17	22	14	3	1
With (±) differences of							
10.0 percent or more	148	42	12	26	28	18	22
15.0 percent or more	117	36	8	17	19	17	20
25.0 percent or more	68	26	5	8	10	5	14
Areas with 500 to 2,499 Population							
Average percent difference	10.1	19.9	7.8	5.6	8.1	11.5	21.9
Number of areas—TOTAL	195	2	24	50	61	29	29
With positive differences	75	2	18	26	18	7	4
With (±) differences of							
10.0 percent or more	64	1	6	9	17	13	18

15.0 percent or more	36	1	3	4	6	6	18
25.0 percent or more	15	1	1	0	0	2	11
Areas with 2,500 to 9,999 Population							
Average percent difference	7.7	13.6	6.7	5.3	5.1	11.3	26.4
Number of areas—TOTAL	199	1	29	72	57	27	13
With positive differences	83	1	22	32	21	5	2
With (±) differences of							
10.0 percent or more	46	1	7	8	7	13	10
15.0 percent or more	32	0	5	4	5	8	10
25.0 percent or more	9	0	1	1	1	0	6
Areas with 10,000 to 24,999 Population							
Average percent difference	5.2	5.5	4.3	3.2	4.9	6.0	18.8
Number of areas—TOTAL	94	2	16	25	32	15	4
With positive differences	42	1	8	13	16	4	0
With (±) differences of							
10.0 percent or more	9	0	1	1	3	2	2
15.0 percent or more	7	0	1	1	3	0	2
25.0 percent or more	3	0	0	0	1	0	2
Areas with 25,000 or More Population							
Average percent difference	2.6	3.0	2.4	2.4	2.5	3.4	6.6
Number of areas—TOTAL	88	2	21	29	25	9	2
With positive differences	46	1	17	16	8	4	0
With (±) differences of							
10.0 percent or more	3	0	1	1	1	0	0
15.0 percent or more	1	0	0	1	0	0	0
25.0 percent or more	0	0	0	0	0	0	0

[a] Percent difference for each area equals postcensal estimate as of July 1, 1975, *minus* adjusted special census count (interpolated or extrapolated to July 1, 1975), expressed as percent of adjusted census count. Average percent difference calculated as arithmetic mean of percent differences *disregarding sign.*

SOURCE: Unpublished data from the Bureau of the Census provided by Frederick Cavanaugh.

TABLE 2.9   Standardized Average Percent Difference Between 1975 Estimates of Population and Special Census Counts, by 1970 Population and Percent Change in Population, 1970–1975: 799 Subcounty Areas With Special Censuses Taken During 1975

1970 Population and Percent Change in Population, 1970–1975	Percent Distribution of Subcounty Areas		Average Percent Difference	
	With 1975 Special Census	All Areas	Not Standard-ized[a]	Standard-ized[b]
All Subcounty Areas	100.0	100.0	11.7	12.3
By 1970 Population				
Less than 1,000	37.5	55.8	19.5	17.0
1,000 to 4,999	28.7	29.0	8.8	7.2
5,000 to 9,999	11.0	6.4	8.2	7.0
10,000 to 49,999	18.0	7.3	4.3	3.8
50,000 or more	4.8	1.5	2.6	2.6
By Percent Change, 1970–1975				
−10.0 percent or more	7.1	9.1	38.8	38.1
−0.0 to −9.9 percent	14.4	24.4	7.3	9.3
+0.0 to +4.9 percent	14.3	19.4	6.4	7.9
+5.0 to +9.9 percent	13.0	15.3	6.9	9.0
+10.0 to +49.9 percent	41.7	30.0	8.9	10.8
+50.0 percent or more	9.5	1.7	24.1	24.4

[a] Average percent differences for 799 subcounty areas in which special censuses were taken in 1975, calculated as in Tables 2.7 and 2.8.

[b] Average percent differences calculated by reweighting the averages using the "size and percent change in size" composition of all subcounty areas for which population estimates are made by the Census Bureau. Thus the average percent difference for "all subcounty areas" is reweighted by the cross-classified "population size by percent change in population" composition of all subcounty areas for which estimates were made. Similarly, the average percent difference for each "population size" group is reweighted by the "percent change in population" composition of all subcounty areas in that size group. And the average percent difference for each "percent change" group is reweighted by the population size composition of all subcounty areas in that "percent change" group.

SOURCE: Unpublished data from the Bureau of the Census provided by Meyer Zitter and Frederick Cavanaugh.

county areas eligible for GRS funds. It is also known that special census areas are selective of fast-growing areas. If we reweight (that is, standardize) the overall average percent difference to reflect the distribution of all subcounty areas by population size in 1970 and percent change in population, for 1970–1975, we obtain a standardized overall average of 12.3 percent (Table 2.9). The standardization assigns more weight to small areas (which tend to have large errors) and to slowly changing areas (which tend to have small errors).[5] Reweighted average percent differences were smaller for areas in each population-size group except in that with 25,000 or more population. For example, the unstandardized average percent difference for subcounty areas with less than 1,000 population was 19.5 as compared with a reweighted average percent difference of 17.0. This difference arises because the areas in which special censuses were taken contain a larger proportion of fast-growing areas than all subcounty areas for which population estimates are made, and fast-growing areas are subject to larger error than other areas. In general, the reweighting has little impact on the overall average error and on the pattern of error by population size and by rate of change in population.

As in the case of the county estimates, there is strong evidence of bias in the subcounty estimates. The estimation method consistently tends to underestimate the population of growing areas and to overestimate the population of declining areas. This can be seen in the third column of Table 2.7, which reports the proportion of differences between estimates and special census counts that were positive (i.e., overestimates). For example, 84 percent of the estimates for areas that declined in population by 10 percent or more between 1970 and 1975 were overestimates, as were 72 percent of the estimates for areas that declined by less than 10 percent. Similarly, 91 percent of the estimates for areas that had grown by 50 percent or more were underestimates, as were 78 percent of the estimates for places that had grown by 25 to 49 percent.

The low levels of accuracy of the estimates for small areas, and for areas undergoing rapid growth or decline, are evident in the measures of extreme error in the last three columns of Table 2.7. Among areas with less than 500 population, two-thirds (66 percent) had differences between population estimates and census counts of at least 10 percent, more than one-half (52 percent) had differences exceeding 15 percent, and almost

---

[5]The Census Bureau provided the Panel with a cross-tabulation of GRS areas by size of population (1970) and percent change in population (1970–1975). Standardization for 1970 population size alone increased the average difference from 11.7 to 14.3 percent; standardization for 1970–1975 change in population alone decreased the average difference to 10.7 percent.

one-third (30 percent) had differences that exceeded 25 percent. Among areas with 500 to 2,499 population in 1970, one-third had differences of 10 percent or more and almost one-fifth had differences of 15 percent or more.

Even more striking are the measures of extreme error for subcounty areas that experienced rapid population growth or that declined in population between 1970 and 1975. Over three-fourths of the areas that declined by 10 percent or more had errors of at least 10 percent, and almost one-half had errors of at least 25 percent. Similarly, of the areas that grew by 50 percent or more, two-thirds had errors of at least 15 percent, and 43 percent had errors of 25 percent or more. Among areas that grew by 25 to 49 percent, 30 percent had errors of 15 percent or more. (The detailed distributions by size and direction of percent error are reported in Tables 2.10 and 2.11 for subcounty areas classified by population in 1970 and by percent change in populations 1970–1975.)

It should be noted that estimates for counties are considerably more accurate than estimates for subcounty areas of the same size and rate of change in population. For example, counties with 1,000 to 4,999 population had an average percent difference of only 5.2 as compared with a difference of 8.8 percent for subcounty areas of the same size (see Tables 2.1 and 2.9).

Thus far our evaluation of the accuracy of the population estimation methods has been based on percent differences between the estimates of total population and special census counts. The Census Bureau's own evaluations of their estimates have been based on similar measures (see, for example, Bureau of the Census (1973b, 1980)). Two considerations suggest, however, that the estimation methods should also be evaluated in terms of the accuracy with which they measure change in population since the last decennial census. First, the methods are designed to measure change in population since the last census: estimates of total population are produced by adding the estimated change in population to the previous census counts. Second, the usefulness of the estimates as updates for the purpose of allocating general revenue sharing funds between regular censuses depends on the accuracy of the estimated changes in population. If the estimated change in population for a substantial number of areas is in the wrong direction, or if the average error of the estimated change is excessively large, it may be preferable to use previous census counts for allocation purposes.

Therefore it is worth noting that percent differences between estimated change and enumerated change in population would be much larger than the percent differences between total population estimates and enumerated census counts that are summarized in Tables 2.7–2.11. Moreover, the pattern of differences would be substantially altered, since the same difference in number between an estimate and the comparable special census count would be expressed as a percent of the change in population rather than as a percent of the total population of the area.[6] The tremendous impact such a shift in base would have on measures of percent error is demonstrated in the illustrative calculations below, which convert the average percent differences based on total population (from Table 2.8) to average percent differences based on change in population. The following table gives the average percent differences between postcensal estimates and special census figures for 1975:

	Percent Change in Population, 1970-1975					
1970 Population	−10.0 or More	−0.0 to −9.9	+0.0 to +9.9	+10.0 to +24.9	+25.0 to +49.9	+50.0 or More
Less than 500						
based on change	243	260	317	91	73	64
(based on total)	(42.9)	(13.7)	(15.1)	(13.5)	(20.0)	(27.4)
500 to 2,499						
based on change	113	148	118	54	42	51
(based on total)	(19.9)	(7.8)	(5.6)	(8.1)	(11.5)	(21.9)
2,500 to 9,999						
based on change	77	127	111	34	41	62
(based on total)	(13.6)	(6.7)	(5.3)	(5.1)	(11.3)	(26.4)
10,000 to 24,999						
based on change	31	82	67	33	22	44
(based on total)	(5.5)	(4.3)	(3.2)	(4.9)	(6.0)	(18.8)
25,000 or more						
based·on change	17	46	50	17	12	15
(based on total)	(3.0)	(2.4)	(2.4)	(2.5)	(3.4)	(6.6)

[6] Algebraically, if C = 1975 special census count (adjusted), E = 1975 population estimate, P = 1970 population, then $|(E - C)/C|100$ equals the percent difference between estimate of total population and total census count (disregarding sign) and $|[(E - P) - (C - P)]/(C - P)|100 = |(E - C)/(C - P)|100$, which equals the percent difference between estimated and enumerated change in population (disregarding sign). Note also that

$$\left| \frac{E - C}{C - P} \right| 100 = \left| \frac{\dfrac{E - C}{C}100}{\left(\dfrac{C - P}{P}100\right)\dfrac{P}{C}} \right| 100$$

$$= \left| \frac{\text{percent difference for total estimate}}{\left(\begin{array}{c}\text{percent change in} \\ \text{population, 1970-1975}\end{array}\right)\left(\dfrac{\text{1970 population}}{\text{1975 count}}\right)} \right| 100.$$

TABLE 2.10 Distribution of Subcounty Areas by Size of Percent Difference Between 1975 Population Estimates and Special Census Counts, by 1970 Population and Percent Change in Population, 1970–1975: 799 Subcounty Areas With Special Censuses Taken During 1975

1970 Population and Percent Change in Population, 1970–1975	Total	Size of Percent Difference[a]										
		−25.0 or More	−15.0 to −24.9	−10.0 to −14.9	−5.0 to −9.9	−0.0 to −4.9	Exact 0	+0.0 to +4.9	+5.0 to +9.9	+10.0 to +14.9	+15.0 to +24.9	+25.0 or More
All subcounty areas	799	49	57	43	116	179	11	163	60	32	43	46
By 1970 Population												
Less than 500	223	31	28	19	20	17	10	16	12	12	21	37
500 to 999	77	4	5	6	18	10	1	14	9	6	3	1
1,000 to 2,499	118	8	12	10	26	20	0	20	13	4	3	2
2,500 to 4,999	111	2	6	3	22	35	0	24	7	3	6	3

(cut off)	2	3	4	15	22	0	22	6	4	6	2
10,000 to 24,999	94	2	1	1	9	39	0	27	10	1	3	1
25,000 to 49,999	50	0	0	0	4	19	0	24	2	0	1	0
50,000 or more	38	0	0	0	2	17	0	16	1	2	0	0

By Percent Change, 1970–1975

−10.0 percent or more	57	2	2	1	2	2	2	3	4	6	8	25
−0.0 to −9.9 percent	115	1	2	3	2	24	1	43	18	7	8	6
+0.0 to +4.9 percent	114	2	1	4	14	23	4	41	11	2	8	4
+5.0 to +9.9 percent	104	0	4	4	13	40	0	18	9	8	5	3
+10.0 to +24.9 percent	228	8	13	17	44	67	2	44	15	6	8	2
+25.0 to +49.9 percent	105	5	19	12	28	18	0	11	2	3	5	2
+50.0 percent or more	76	31	16	2	13	5	2	3	1	0	1	2

[a] Percent difference for each county equals postcensal estimate (as of July 1) *minus* adjusted special census count (interpolated or extrapolated to July 1), expressed as percent of adjusted census count.

SOURCE: Unpublished data from the Bureau of the Census provided by Frederick Cavanaugh.

TABLE 2.11 Distribution of Subcounty Areas by Size of Percent Difference Between 1975 Population Estimates and Special Census Counts, by 1970 Population Cross-Classified by Percent Change in Population, 1970-1975: 799 Subcounty Areas With Special Censuses Taken During 1975

Size of Percent Difference[a] and 1970 Population	Total	Percent Change in Population, 1970-1975					
		−10.0 or More	−0.0 to −9.9	+0.0 to +9.9	+10.0 to +24.9	+25.0 to +49.9	+50.0 or More
Less than 500 population	223	50	25	42	53	25	28
−25.0 percent or more	31	2	1	2	8	4	14
−15.0 to −24.9 percent	28	2	1	4	6	10	5
−10.0 to −14.9 percent	19	1	2	6	7	1	2
−5.0 to −9.9 percent	20	2	0	2	8	4	4
−0.0 to −4.9 percent	17	0	3	3	8	3	0
Exact 0	10	2	1	3	2	0	2
+0.0 to +4.9 percent	16	3	6	3	4	0	0
+5.0 to +9.9 percent	12	1	3	5	3	0	0
+10.0 to +14.9 percent	12	5	2	3	2	0	0
+15.0 to +24.9 percent	21	8	2	5	3	2	1
+25.0 percent or more	37	24	4	6	2	1	0
500 to 2,499 population	195	2	24	50	61	29	29
−25.0 percent or more	12	0	0	0	0	1	11
−15.0 to −24.9 percent	17	0	0	1	5	4	7
−10.0 to −14.9 percent	16	0	1	2	8	5	0
−5.0 to −9.9 percent	44	0	2	7	20	10	5
−0.0 to −4.9 percent	30	0	3	13	10	2	2
Exact 0	1	0	0	1	0	0	0
+0.0 to +4.9 percent	34	0	6	13	8	4	3
+5.0 to +9.9 percent	22	1	7	7	6	0	1
+10.0 to +14.9 percent	10	0	2	3	3	2	0
+15.0 to +24.9 percent	6	0	2	3	1	0	0
+25.0 percent or more	3	1	1	0	0	1	0
2,500 to 9,999 population	199	1	29	72	57	27	13
−25.0 percent or more	4	0	0	0	0	0	4
−15.0 to −24.9 percent	11	0	0	0	2	5	4
−10.0 to −14.9 percent	7	0	0	0	2	5	0
−5.0 to −9.9 percent	37	0	0	15	12	8	2
−0.0 to −4.9 percent	57	0	7	25	20	4	1
Exact 0	0	0	0	0	0	0	0
+0.0 to +4.9 percent	46	0	12	17	16	1	0
+5.0 to +9.9 percent	13	0	3	7	2	1	0
+10.0 to +14.9 percent	7	1	2	4	0	0	0
+15.0 to +24.9 percent	12	0	4	3	2	3	0
+25.0 percent or more	5	0	1	1	1	0	2

TABLE 2.11 *Continued*

Size of Percent Difference[a] and 1970 Population	Total	Percent Change in Population, 1970–1975					
		−10.0 or More	−0.0 to −9.9	+0.0 to +9.9	+10.0 to +24.9	+25.0 to +49.9	+50.0 or More
10,000 to 24,999 population	94	2	16	25	32	15	4
−25.0 percent or more	2	0	0	0	0	0	2
−15.0 to −24.9 percent	1	0	1	0	0	0	0
−10.0 to −14.9 percent	1	0	0	0	0	1	0
−5.0 to −9.9 percent	9	0	0	2	2	5	0
−0.0 to −4.9 percent	39	1	7	10	14	5	2
Exact 0	0	0	0	0	0	0	0
+0.0 to +4.9 percent	27	0	4	11	9	3	0
+5.0 to +9.9 percent	10	1	4	1	4	0	0
+10.0 to +14.9 percent	1	0	0	0	0	1	0
+15.0 to +24.9 percent	3	0	0	1	2	0	0
+25.0 percent or more	1	0	0	0	1	0	0
25,000 or more population	88	2	21	29	25	9	2
−25.0 percent or more	0	0	0	0	0	0	0
−15.0 to −24.9 percent	0	0	0	0	0	0	0
−10.0 to −14.9 percent	0	0	0	0	0	0	0
−5.0 to −9.9 percent	6	0	0	1	2	1	2
−0.0 to −4.9 percent	36	1	4	12	15	4	0
Exact 0	0	0	0	0	0	0	0
+0.0 to +4.9 percent	40	0	15	15	7	3	0
+5.0 to +9.9 percent	3	1	1	0	0	1	0
+10.0 to +14.9 percent	2	0	1	0	1	0	0
+15.0 to +24.9 percent	1	0	0	1	0	0	0
+25.0 percent or more	0	0	0	0	0	0	0

[a] Percent difference for each area equals postcensal estimate as of July 1 *minus* adjusted special census count (interpolated or extrapolated to July 1), expressed as percent of adjusted census count.

SOURCE: Unpublished data from the Bureau of the Census provided by Frederick Cavanaugh.

The above calculations were made by assuming that the percent change in population for all subcounty areas in each size-percent change subgroup in Table 2.8 was exactly the midpoint of the percent change interval; it was also assumed that all areas that declined in population by 10 percent or more had declined by exactly 15 percent and that all areas that increased in population by 50 percent or more increased by exactly 75 percent. For example, the percent difference based on change in population for areas with 500 to 2,499 population that increased by 0.0 to 9.9 percent between 1970 and 1975 was calculated under the assumption that each of the 50 areas in this subgroup had increased by exactly 5.0 percent. Thus the percent difference between the estimated and enumerated change in population was calculated from the average percent difference of 5.6 percent (based on total population) for this subgroup of areas in Table 2.8 using the formula in footnote 6 above, namely,

$$
\begin{array}{l}\text{percent difference}\\ \text{based on change in}\\ \text{population}\end{array} = \frac{\text{percent difference based on total population}}{\left(\begin{array}{c}\text{percent change in}\\ \text{population, 1970–1975}\end{array}\right)\left(\dfrac{\text{1970 population}}{\text{1975 count}}\right)}
$$

$$
= \frac{5.6}{(5.0)\left(\dfrac{1.00}{1.05}\right)} = \frac{5.6}{4.76} = 118.
$$

The patterns of error in the illustrative calculations are strikingly different from those in Table 2.8. Subcounty areas subject to little growth or decline have the largest percent differences based on change in population, whereas the fast-growing areas have much smaller percent differences. From this perspective the greater accuracy documented in Tables 2.7–2.11 for areas of slow or moderate change in population size is explained by the fact that their change in population from 1970 to 1975 was a smaller proportion of their total population in 1975 than was the case for areas undergoing more rapid rates of growth or decline.

It is also worth noting how very large the relative errors are when they are based on change in population rather than on total population. For example, the rather moderate average percent difference of 5.3 (based on total population) for areas with 2,500 to 9,999 population (in 1970) that grew by less than 10 percent between 1970 and 1975 represents an average difference of 111 percent when the error is measured in terms of change in population size (and when it is assumed that all areas increased by exactly 5 percent). Similarly, the average percent difference of 11.3 percent (based on total population) for areas of the same size that grew by 25 to 49 percent between 1970 and 1975 represents an average difference of 41 percent when the error is measured in terms of change in population. Although these illustrative calculations are not based on real data for individual subcounty areas, they are true measures of the percent differences based on change in population under the stated assumption, namely, that all subcounty areas in each subgroup of areas in Table 2.8

had the same percent change in population, the midpoint of the percent change interval.

We have made some real calculations for one population-size group of subcounty areas. Percent differences between estimated and enumerated change in population for 118 subcounty areas that had 1,000 to 2,499 population in 1970 are reported in Table 2.12. The pattern of differences based on change in population is quite similar to the illustrative calculations derived above for areas with 500 to 2,499 population. The more detailed classification by percent change in population for 1970-1975 (nine intervals instead of six) provides us with separate measures for areas with little change in population (less than 5 percent increase or decrease). The average percent differences (based on change in population) for these two groups of areas were exceedingly high—247 and 331 percent—largely because of the very small base of the percent differences for individual areas in these two groups. The average percent differences for areas that increased or decreased by 5 to 10 percent during 1970-1975 were also very high—100 percent or more. Although the average error decreased as the rate of population growth increased, it remained as high as 49 percent for areas that grew by 25 percent or more between 1970 and 1975.

In evaluating the accuracy of estimates of postcensal change in population, it is also important to take into account whether the estimated change is in the correct direction, that is, does it correctly estimate whether the population increased or decreased. The average percent differences in Table 2.12 are averages of unsigned percent differences. Thus an average (based on change in population) that exceeds 100 percent indicates that the error in the estimate of change was larger (on the average) than the enumerated change in population: the estimate of change either was in the wrong direction or overestimated the magnitude of change by more than 100 percent. The third column of Table 2.12 reports the number of subcounty areas for which the percent difference based on change in population exceeded 100 percent, and the fourth column indicates the number of areas in which the estimated change in population was in the wrong direction (increase instead of decrease or vice versa). In 20 of the 118 areas the estimated change was in the wrong direction. In 10 areas that actually decreased in population according to special census counts, the estimates showed an increase in population; most (8 of 10) of these areas had declined in population by less than 5 percent, but the estimated increase for 3 of these 8 areas exceeded 5 percent. Similarly, in another 10 areas that actually increased in population, the estimates showed a decrease in population; almost half (4 of 10) of these areas had actually increased by 15 to 24 percent although the estimates showed

TABLE 2.12   Percent Difference Between Estimated and Enumerated Change in Population, 1970–1975, by Percent Change in Population, 1970–1975: 118 Subcounty Areas With 1,000 to 2,499 Population in 1970 With Special Censuses Taken in 1975

Percent Change in Population, 1970–1975	Average Percent Difference Based on Change[a]	Total Number of Areas	Number of Areas with Percent Difference Greater Than 100.0[b]		Average Percent Difference Based on Total[c]
			Total	Estimate of Change in Wrong Direction	
−10.0 or more	140	2	1	1	19.9
−5.0 to −9.9	114	4	2	1	7.7
−0.0 to −4.9	331	11	8	8	4.7
+0.0 to +4.9	247	13	6	2	3.4
+5.0 to +9.9	100	13	4	2	6.4
+10.0 to +14.9	64	17	3	2	7.2
+15.0 to +24.9	63	19	4	4	10.7
+25.0 to +49.9	49	20	1	0	12.2
+50.0 or more	49	19	0	0	20.8
TOTAL	111	118	29	20	10.3
Total (excluding −4.9 to +4.9 percent change)	87	103	14	10	10.9

[a] Percent difference for each area equals estimated change in population, 1970–1975 (1975 estimate *minus* 1970 population) *minus* enumerated change in population (1975 adjusted census count *minus* 1970 population), expressed as percent of enumerated change in population. Average calculated as arithmetic mean of percent differences *disregarding sign*.
[b] Refers to percent difference based on change in population.
[c] Percent difference for each area equals 1975 postcensal estimate *minus* adjusted special census count (interpolated or extrapolated to July 1), expressed as percent of adjusted census count. Average calculated as arithmetic mean of percent differences *disregarding sign*.

SOURCE: Unpublished data from the Bureau of the Census provided by Frederick Cavanaugh.

decreases of 1 to 9 percent. In addition to the 20 areas for which the estimated change was in the wrong direction, there were 9 subcounty areas for which the change in population was overestimated by more than 100 percent (8 were overestimates of population increase, and 1 was an overestimate of population decline).

These individual calculations for 118 subcounty areas, together with the illustrative calculations reported earlier, raise questions about the advisability of attempting to update population data for purposes of allocating funds to small areas. It seems quite possible that the estimated postcensal change in population for a substantial proportion of areas below some as yet unspecified threshold may be in the wrong direction or may have average errors in excess of 100 percent. This possibility should be carefully checked in tests of the estimation methods against the 1980 census results.

It is also probable that factors other than population size and rate of population change—for example, age structure of the population—affect the accuracy of the subcounty estimates. These relationships should be further explored when the 1980 census results are available.

2.3 PANEL EVALUATION OF PER CAPITA INCOME ESTIMATES

Tests of per capita income estimates are performed with two considerations in mind: (1) accuracy of the per capita income estimates as used for GRS and (2) accuracy of the estimates of postcensal change in per capita income. The second consideration is relevant for evaluating the estimation methodology for postcensal per capita income. The basis of our evaluation is a sample of 86 special censuses, taken in 1973 at the Census Bureau's expense, in which income questions were asked of the entire enumerated populations.

Although the same tests are performed to evaluate accuracy for points 1 and 2, the test results are interpreted differently. As was noted above, the postcensal estimate of per capita income level for an area equals the sum of the 1970 census estimate and the estimate of postcensal change. Since the 1970 census estimates are based on 20-percent samples of respondents, they are subject to sampling error. Thus the estimates of postcensal level contain error from the estimation of change, and they also contain sampling error from the 1970 estimates. The effect of the latter error needs to be eliminated when one makes inferences about point 2.

Concern over accuracy of GRS allocations leads us to focus on accurate estimation of the ratio of subcounty (or county) per capita income to county (or state) per capita income.[7] If the per capita income for every place in a county is underestimated by the same proportion, then so is the per capita income of the county, and these errors cancel so as to cause no error in the subcounty GRS allocations. Thus uniform proportional errors would not be important. Unfortunately, the 86 special censuses provide insufficient data for us to evaluate successfully the differential errors in the per capita income estimates: by differential error we mean differences between the proportional errors of subcounty units in the same county (or counties in the same state).

Errors in the 86 special census data on per capita income also hinder our attempts at evaluation of accuracy for points 1 or 2. Although these data are not based on sampling but on attempted complete enumeration, nonresponse to questions on income and biased response (e.g., under-reporting of income) both introduce error. Previous studies have indicated that income was underreported in the 1970 census by about 8 percent for the nation as a whole (Ono, 1972). The underreporting varied significantly by type of income (wage and salary income, farm income, etc.), and dif-ferential errors among places were substantial. Underreporting and nonresponse undoubtedly also mar the special census figures, but we can only guess at the extent.[8]

Thus the special census data on per capita income are inaccurate to some unknown degree, and the difference between the special census figure and the postcensal estimate of per capita income for a place may not be caused entirely by error in the latter. Alternatively, the errors in the two figures can conceivably offset each other, so that their difference may on occasion underestimate the error in the postcensal estimate.

Our comparison of the results of the 1973 special censuses to the post-censal estimates for the same date is shown in Table 2.13. The second col-umn shows the percentage of areas for which the postcensal estimate was closer to the special census figure than was the 1970 census figure. It should be noted—the third and fourth columns—that the postcensal esti-mates of level for the smallest places (1970 population under 1,000) are not much better estimates of level than are the 3-year-old 1970 census estimates, despite inflation. Since inflation causes the per capita income levels to rise more or less uniformly for most places, much of the discrepancy between the 1970 census and 1973 special census figures

---

[7]The denominators given here are approximate; see Appendix E for the actual, more com-plicated expressions.

[8]For the special censuses, nonrespondents were assumed to have the same income as respondents. For the 1970 census, more sophisticated imputation techniques were used. Fay and Herriot (1979) suggest that those techniques cause a relative downward bias in the special census estimates.

TABLE 2.13 Comparison of 1973 Special Census Per Capita Income Estimates With 1970 Estimates and Postcensal Estimates: Original Methodology

1970 Population of Places	Number of Places	Percent of Areas for Which 1973 Postcensal Estimate Closer Than 1970 Census to 1973 Special Census	Average Percent Difference From 1973 Special Census[a]	
			1970 Census	1973 Estimate
Under 500	16	62	25	28
500–999	11	45	15	17
1,000–4,999	46	65	15	10
5,000–9,999	9	89	15	8
10,000–20,000	4	100	21	14
TOTAL	86	66	17	15
Total above 500	70	67	15	11
Total above 1,000	59	71	15	10

[a] Percent difference for each place equals postcensal estimate minus special count, expressed as percent of census count. Average percent difference calculated as arithmetic mean of percent differences *disregarding sign.*

SOURCE: Unpublished data from the Bureau of the Census provided by Roger Herriot.

would disappear if ratios of income levels were the focus of comparison. We note, for example, that for only 9 of the 86 areas did the 1973 census per capita income figure fall below the 1970 census figure. We suspect that the postcensal estimates would look even worse if we could similarly compare the ratios of place per capita income to county per capita income. As we mentioned above, for general revenue sharing, ratios rather than levels of per capita income are relevant (see Appendix E).

The methodology underlying the postcensal per capita income estimates analyzed in Table 2.13 was modified later in the 1970s. The Census Bureau originally estimated per capita income for subcounty units with 1970 population below 500 by the estimated per capita income for the whole county; for subcounty units with 1970 population of 500 to 999 the Bureau estimated per capita income by attributing to these units the estimated rate of change for the aggregate of all areas in the county with under 10,000 population in 1970. Beginning with the estimates of per capita income for 1974, empirical Bayes techniques and other modifications were used to revise the procedures for these very small places (population under 1,000). Fay and Herriot (1979, Table 3) recomputed estimates for 24 of the 86 units under the revised methodology. For 16 places with 1970 population under 500, the average difference between the special census and the 1973 special census estimate decreased from 28 to 22 percent; for 8 places with population between 500 and 1,000 the corresponding difference decreased from 21 to 17 percent.[9] The proportion of places for which the 1973 estimate was closer than the 1970 estimate was to the special census was unchanged. The revised methodology apparently improves the accuracy of the postcensal estimates. Other tests (the "Groups of Ten Test" discussed by Fay and Herriot (1979) (section 4) and the Bureau of the Census (1980)) show that the revised methodology improves the 1970 base estimates as well. Nevertheless, we conclude that for GRS purposes the use of postcensal estimates of per capita income for the smallest places may not be substantially more accurate than the use of 1970 census estimates (especially if the latter are adjusted by empirical Bayes techniques), although this conclusion might not hold for longer time periods.

In evaluating the estimates of postcensal change in per capita income, as was mentioned above, the comparisons of postcensal estimates of change with censal estimates of change[10] are confounded by underreporting errors and nonresponse errors. It is sometimes hypothesized that the errors arising from underreporting of income are stable over time so that these errors cancel when one considers changes in the estimates over time. For example, if underreporting caused per capita income for an area to be underestimated by $200 both by the 1970 census and by the 1973 special census, then the errors cancel, and the difference between the two estimates accurately measures the true change in per capita income. In this case the difference between the 1970 census and 1973 special census estimates would be a good standard for assessing the accuracy of the updates. However, such neat cancellation of underreporting and nonresponse errors may be more hoped for than real (see Appendix I).

In addition to underreporting and nonresponse errors, sampling error contributes to the inaccuracy of the difference between 1973 special census and 1970 census estimates as an estimate of postcensal change in per

---

[9] Fay and Herriot (1979, Table 3) classify areas by the 1970 census weighted sample population rather than the 1970 census count, as we do. The classifications are the same for all areas except Bonaparte, Iowa, which had a 1970 population of 517 but a weighted sample population under 500.

[10] That is, comparison of the difference of the postcensal estimate minus the 1970 census estimate with the difference of the 1973 special census estimate minus the 1970 census estimate.

capita income.[11] Unpublished calculations by the Census Bureau indicate that the coefficient of variation due to sampling for an area was approximately 3.0 divided by the square root of the size of the area's 1970 population. For small areas the coefficient of variation is large: .09 for an area with 1,000 population and .30 for an area with 100. The presence of these errors implies that one cannot simply interpret deviations between the "census change" (the difference between the 1973 special census estimate and the 1970 census estimate) and the estimated change (the difference between the postcensal estimate and the 1970 census estimate) as evidence of error in the estimate of postcensal change. To estimate this error, it would be necessary to separate out the other components of error—nonresponse, underreporting, and sampling error in the census estimates.[12] We do not undertake this task here, but possible approaches are noted in section 3.3.

---

[11] Although the 1970 census estimate enters into both of the quantities being compared, the difference between the postcensal estimate of level and the 1970 census estimate is the actual estimate of change, but the difference between the 1973 special census estimate and the 1970 census estimate is an inaccurate estimate of true change because of sampling error in the 1970 census estimates.

[12] An extensive bibliography on error components is given by Sahai (1979).

3 Alternative Approaches to Evaluation

3.1 LOSS FUNCTIONS AND OPTIMIZATION CRITERIA

Chapter 1 (section 1.1d) discussed four criteria of accuracy along with the likelihood of conflicts among them. The four criteria are (1) low average error, (2) low average relative error, (3) few extreme relative errors, and (4) absence of bias for subgroups.

An explicit, concise, and useful way to summarize weightings of accuracy criteria is to formulate loss functions or optimization criteria. These devices are designed so that choosing a procedure to minimize them corresponds to selecting an estimation procedure best satisfying the accuracy criteria and the preferred trade-offs among them. For example, a familiar optimization criterion for estimating a single parameter is mean square error: one chooses the estimator with the smallest mean square error.

Before giving illustrations of loss functions and optimization criteria, we note two applications for small-area estimation. One application is for selecting the "best" from a class of alternative estimators for which data are available. Consider, for example, choosing among weighted averages of two estimators, one having low average relative error but some extreme errors and the other with higher average relative error but no extreme errors. One can choose the weightings in the average so as to minimize a specified optimization criterion. This is a familiar statistical problem.

A second application relates to collecting data and designing estimators for which the data must be gathered. This use of optimization criteria is more difficult than the first because the costs of producing estimates (data collection, analysis, etc.) must be considered. This problem has been carefully considered for some special situations: an example is the determination of optimal sample designs, such as the "Neyman allocation" for stratified sampling. Estimators for postcensal population and income are not usually based directly on sample data, and the costs of providing the necessary data are essentially different from the costs of sample data. For instance, sample estimates with given coefficients of variation for small and large places may cost approximately as much for small places as for large places. But AR method estimates, based on tax returns, may be more expensive for small places than for large places if coefficients of variation are required to be equal. The extra cost for small places arises because boundary changes and geographic coding problems are generally more important for small places than for large places.

Let P_i and $\hat{P}_i$ denote the actual[1] and estimated population of the ith local area, for a total of n local areas. Consider the second criterion, low average relative error. Attaining that criterion is equivalent to minimizing

$$\sum_i |\hat{P}_i - P_i|/P_i. \tag{3.1}$$

To reflect the third accuracy criterion, few extreme errors (or low variation in error), optimization criterion (3.1) can be modified to

$$\sum_i (|\hat{P}_i - P_i|/P_i)^a \tag{3.2}$$

where a is a number larger than 1.[2]

Large values of a reflect a desire to reduce extreme errors. For example, consider choosing between X and Y, two alternative sets of estimates for n places of approximately the same size. Suppose X had equal relative absolute errors of .04 for all places and Y had a relative error of .20 for 1 percent of the places and .01 for 99 percent of the places. In this case the average relative error for Y (.0119) is less than the average relative error for X (.04). But if criterion 3 is used with a greater than or equal to 2.85, set X will be selected. As a grows large without bound, minimization of

---

[1] In practice, when one is estimating the value of (3.1) below, the value of P_i is not known but is estimated, often on the basis of a census or survey; care is needed to adjust for error in this estimate (see Appendix I).

[2] Since the errors are random, minimization of (3.1) or (3.2) refers to minimization of the expected value of (3.1) or (3.2) or of some strictly increasing transformation of (3.1). For example, if $a = 2$ in (3.2), one might minimize the expectation of the square root of (3.2), often referred to as the root-mean-square error.

criterion 3 becomes equivalent to minimization of the largest relative error (i.e., minimization in expected value). Intermediate values of a can reflect trade-offs between the criteria of low average relative error and of few extreme errors.

Now consider the problem of designing or selecting an estimator to minimize errors in allocations of funds. "To minimize errors in allocations" is a vague statement that allows several interpretations. For example, one could seek to minimize the total number of dollars misallocated (i.e., allocated to the wrong area). For the ith local area, let $\hat{A}_i$ and A_i denote the estimated allocation and the targeted allocation if there are no errors in the data. One seeks to minimize the expectation of

$$\sum_i |\hat{A}_i - A_i|. \tag{3.3}$$

Note that $\hat{A}_i$ and A_i involve $\hat{P}_i$ and P_i, respectively, in an implicit fashion, so that (3.3) is a complicated expression of $\hat{P}_1, \ldots, \hat{P}_n$ and $P_1, \ldots, P_n$. One may not be able to specify exactly how to choose the estimators $\hat{P}_1, \ldots, \hat{P}_n$ to minimize (3.3); in fact, one may not even be able to write out (3.3) explicitly in terms of $\hat{P}_1, \ldots, \hat{P}_n$ and $P_1, \ldots, P_n$, but approximations are possible (see Appendix E). For illustrative purposes we will simplify greatly and assume that $\hat{A}_i$ and A_i are proportional to the fraction of total population living in the ith area; that is, we assume that

$$A_i = cP_i/(\sum_j P_j) \qquad \hat{A}_i = c\hat{P}_i/(\sum_j \hat{P}_j),$$

for some positive constant c not depending on i. In this special case we may rewrite (3.3) as

$$c\sum_i |\hat{P}_i/\sum_j \hat{P}_j - P_i/\sum_j P_j|. \tag{3.4}$$

Note that uniform relative errors are irrelevant; for example, if $(\hat{P}_i - P_i)/P_i$ is the same for all areas i, then (3.4) is zero. Simplifying even further, let us assume that $\sum \hat{P}_j$ estimates $\sum P_j$ with negligible error, so that (3.3) and (3.4) can now be expressed as

$$G\sum_i |\hat{P}_i - P_i|, \tag{3.5}$$

where G is some positive constant of proportionality. Ignoring the constant of proportionality, we notice that (3.1), (3.2), and (3.5) are special cases of the general optimization criterion

$$\sum_i P_i q(|\hat{P}_i - P_i|/P_i)^a \qquad (3.6)$$

for some nonnegative values of q and a.[3]

By appropriately choosing values of a and q one can use (3.6) to reflect compromises or trade-offs among alternative criteria of accuracy. As we observed earlier, setting $q = 0$ and choosing intermediate values of $a > 1$ achieves a trade-off between the criteria of low average relative error and of few extreme errors. Choosing q larger than zero but smaller than a effects a further compromise between the trade-off above and the criterion of minimal errors in allocation of funds. What are appropriate values for a and q is largely a policy question and not a technical question.

The following example illustrates the implications of different values of a and q for selecting estimators for local areas. For clarity of presentation we consider the substate jurisdictions of the United States partitioned into two groups on the basis of 1970 population counts: those with at least 10,000 inhabitants will be called "large," and the rest "small." Also for simplicity we assume that all large places have identical population sizes and that all small places have identical population sizes, and we consider selecting an estimator for the two population sizes. Suppose we have three estimators, E_1, E_2, and E_3, which provide unbiased estimates of population with the following coefficients of variation:[4]

	E_1	E_2	E_3
Small places	.100	.085	.075
Large places	.040	.045	.050

The estimators represent different trade-offs between error for large places and error for small places. For large places, E_1 is best, then E_2, and last E_3. For small places the situation is reversed: E_3 is best, E_2 second best, and E_1 worst. Which estimator is best overall?

As a rough approximation to reality, say there are 32,500 small areas, each with a population of 1,500 and 3,000 large areas, each with a popula-

---

[3] Further generalizations are possible, of course: for example, $p_i q$ in (3.6) could be replaced by a more general term W_i. Alternative formulas are also possible; see Stanford Research Institute (1974) or Ferreira (1978).

[4] The coefficient of variation of an estimate is its standard deviation expressed as a proportion of the quantity (here, population) being estimated. We are also assuming here that the expected absolute values of the relative errors are proportional to the coefficients of variation (as is the case when the values of the estimators follow the normal distribution).

tion of 45,000 (see Table 1.1 for the actual figures). We will consider selecting an estimator according to four alternative optimization criteria:

$$\text{(A)} \quad \sum_i (\hat{P}_i - P_i)^2,$$

$$\text{(B)} \quad \sum_i |\hat{P}_i - P_i|,$$

$$\text{(C)} \quad \sum_i (\hat{P}_i - P_i)^2 / P_i,$$

$$\text{(D)} \quad \sum_i |\hat{P}_i - P_i| / P_i.$$

Criterion D requires minimizing average relative absolute error, while criterion B requires minimizing the total number of dollars misallocated (assuming allocations are proportional to population). Criterion A is a variant of B and is less concerned with small individual misallocations and more concerned with large individual misallocations. Criterion C represents a compromise between A and D. Each of these criteria implies a different ranking of the three estimators in order of preference:[5]

$$\text{criterion A:} \quad E_1, E_2, E_3;$$

$$\text{criterion B:} \quad E_2, E_1, E_3;$$

$$\text{criterion C:} \quad E_2, E_3, E_1;$$

$$\text{criterion D:} \quad E_3, E_2, E_1.$$

Clearly, the different criteria have different implications for "best": by criterion A, estimator E_1 is best; by both criteria B and C, estimator E_2 is

---

[5] The numerical values of criteria A-D are given by the following tabulation, for estimators E_1, E_2, and E_3, where $a = 10^9$, $b = (10^6)s$, $c = 10^5$, and $d = (10^3)s$. The constant s is the ratio of the expected absolute value of the relative error to the coefficient of variation; for errors following the normal distribution, s is approximately .8:

	E_1	E_2	E_3
Criterion A	10.5a	12.8a	15.7a
Criterion B	10.3b	10.2b	10.6b
Criterion C	7.0c	6.3c	6.5c
Criterion D	3.4d	2.9d	2.8d

It should be noted that comparisons between numerical values under *different* criteria are not meaningful. For example, if all the values for criterion C were multiplied by 10^5, it would not affect the preference ordering represented by criterion C, but the values would be larger than any others in the tabulation.

best; and by criterion D, estimator E_3 is best. It is interesting to note that the criteria discriminate among different trade-offs in the estimators. For example, both criteria B and C favor an increase of .005 relative absolute error for large places to get a decrease of .015 relative absolute error for small places (E_2 is better than E_1); but criterion C favors and criterion B does not favor an increase of .010 relative error for large places to get a reduction of .025 for small places.

We emphasize that the optimization criteria are meaningful only insofar as they represent the desires of the producer of the estimates (the Census Bureau in this case) for different kinds of accuracy. The above illustration demonstrates that tractable formulations of optimization criteria can be useful for representing preferences for trade-offs in accuracy. Once preferences are stated, a representative optimization criterion can be determined and used for selecting estimators with the desired properties.

Note that the optimization criteria A–D in the example above disregard the issue of bias for subgroups. One way to incorporate concerns about bias into the optimization criteria is to use constraints: only estimators with specified unbiasedness properties may be used. Choice of the "best" estimator within the class of acceptably unbiased ones is then made according to optimization criteria.

Constraining or restricting the class of estimators under consideration is also a useful way to reflect other concerns. Criteria A–D are all related to aggregate error, but one might also be concerned that *no* component error be larger than a specified amount (or percent). A reasonable optimization strategy selects as best only an estimator whose component errors lie within preestablished limits. For example, the consideration of estimators might be restricted to those for which the expected relative absolute error for any place picked at random is less than 0.4. The optimization criteria discussed earlier could then be used to select a best estimator from within this restricted class.

This kind of approach is advocated by Office of Federal Statistical Policy and Standards (1978), which recommends (p. 27):[6]

> That since data errors are inevitable and since statistical resources are limited, priority be given to minimizing the very large errors which may occur in data used for the allocation of funds. ... To the extent that error measurements are available for small geographic areas one should check that relative errors are no greater than a prespecified maximum, but one should not be overconcerned with small errors since their effect on the total distribution is relatively minor.

---

[6] Note the distinction between errors and relative errors.

The use of optimization criteria as described above extends directly to situations involving estimators for several parameters, for example, for population and per capita income.

3.2 USE OF ALTERNATIVE ESTIMATES FOR EVALUATION

Accuracy of postcensal estimates may be evaluated by approaches other than comparisons of postcensal estimates with special census results. This section discusses one such approach, comparisons of postcensal estimates with other estimates; the next section discusses another approach, the use of error models.

For evaluating postcensal estimates we want a procedure that can give a current assessment of the level and direction of error in the estimates, i.e., that can provide levels of error without reference to tests against past censuses. Such a procedure can be devised using up-to-date alternative data or even alternative estimates. Furthermore, in some cases the alternative data can be used to determine the estimates themselves and still provide current estimation of the error. We first describe a general approach and then consider the specific use of sample data.

Let X be the estimator we wish to evaluate (e.g., postcensal estimator) and let Y be an alternative estimator (e.g., one based on sample-survey data). If either X or Y is unbiased, then

$$\text{MSE}(X - Y) = \text{MSE}(X) + \text{MSE}(Y) - 2\,\text{cov}(X, Y) \qquad (3.7)$$

where MSE denotes mean square error and cov denotes covariance. Suppose prior estimates of MSE(Y) and cov(X, Y) are available (in practice, it is often appropriate to assume that cov(X, Y) is zero). Then an estimate of MSE$(X - Y)$ can be constructed from the observed values of X and Y: for example, if MSE$(X - Y)$ is believed to be constant over all observed units, then the average value of $(X - Y)^2$ might be used, and (3.7) is easily solved to yield an estimate of MSE(X). It is important to note that if MSE(Y) is large, then the estimate of MSE(X) can be poor (or even negative!). Note that the assumption of unbiasedness for either X or Y is essential for (3.7) to hold.

Fay (1979) describes two exemplary evaluations conducted along these lines by the Census Bureau. The first uses sample estimates from the 1976 Survey of Income and Education to evaluate regression estimates for the number of children in poverty. The second uses sample estimates of per capita income to evaluate regression estimates for the per capita income of small areas. This second application is presented in more detail by Fay and Herriot (1979) and is discussed further in Appendix J.

We now consider two particular approaches to using up-to-date sampling information to provide current estimates of level and direction of error in postcensal estimates. One would be to select a sample of localities and to take a census in each one. These censuses could then be compared to the various estimates (and combinations thereof) for the same areas and the most accurate procedure selected. The Census Bureau did use this approach in 1973 when they took special censuses in a probability sample of 86 areas to evaluate the accuracy of the administrative records method for small subcounty areas. The problem with this procedure is that the sample of local censuses is prohibitively expensive to take for a sufficiently large number of areas to provide a definitive evaluation.

The second approach is to use an existing large, high-quality sample survey. The high quality is necessary to ensure that sample estimates are unbiased, and the large size is necessary so that sample estimates for selected primary sampling units (PSU's) can be computed.[7] Fortunately, such a sample exists for the Current Population Survey (CPS). It takes complete enumerations in 70,000 households each month. Each PSU for the CPS consists of an independent city or county or two or more contiguous counties. The sample estimates computed for these PSU's have been found to be unbiased, and when the estimates are compared with 1970 census counts, the mean relative squared deviation for the PSU estimates (based on data pooled from five consecutive quarterly surveys) was less than .025 (Ericksen, 1975).

These PSU estimates can be taken as dependent variables in regression equations using as independent variables both symptomatic information usually used in population estimates and even alternative population estimates themselves (e.g., CM II estimates and AR method estimates). As long as the error of the sample estimates has no linear trend with relation to the rate of population growth, the resulting regression equation yields an unbiased estimate of what would be computed if population counts of the dependent variable replaced the sample estimates. The one noticeable point of difference is in the size of the correlation coefficients: because a major component of the variance of the dependent variable is random sampling error, the observed correlations, but not regression coefficients, are shrunk.

[7]Large can either mean high sampling rates for fewer places or low sampling rates for many places; see Appendix H. Unbiased is here taken to mean that the expected values of the sample-based estimates are the same as those of estimates based on a census: that is, undercount is ignored.

The logic of the procedure can be described briefly. Suppose one has k independent variables under consideration. The objective is to use the sample data to derive regression estimates based on the k variables (or a subset of them) with minimum mean square error and then to obtain estimates of the mean square error. For all subsets of the variables containing 1, 2, ..., k variables the multiple correlations with the sample PSU estimates are calculated.[8] Among sets containing a given number of variables, that set is selected for which the squared multiple correlation is largest. In this manner, k sets of variables are selected: a one-variable set, a two-variable set, ..., the k-variable set. The best candidate among the k sets selected is the one yielding regression estimates with the smallest mean square error. Details for estimating the mean square error are given by Gonzalez and Hoza (1978) and Ericksen (1974). Briefly, one obtains the mean squared difference between the regression and sample estimates and subtracts an estimate of the mean squared error of the sample estimates. The resulting difference, modified by appropriate constants, provides an estimate of the mean squared error of the regression estimates.

The computed mean square error can be used to estimate the actual level of error, with two caveats. The first is that the estimate of mean square error is dependent on accurate estimation of the mean square error of the PSU sample estimates (in particular, the within-PSU sampling error), which is not always easy. The second is that the PSU's in the CSP are not generally counties, and the levels of error for the two types of units may differ.

Notice that the sample data can be used in two ways: to determine the regression estimates and to estimate the mean square error of the estimates. If the independent variables in the regression are alternative population estimates, the regression coefficients may be interpreted as optimal weightings for averaging (see section 5.2c).

Sample data can also be used as (low precision) benchmarks against which the postcensal estimates can be directly compared (see Appendix H). For example, subsets of PSU's could be formed by stratifying according to characteristics such as size, rate of growth, or age structure of the population. When the average difference over a subset between the postcensal estimates and the sample estimates is larger than sampling error alone can explain, there is evidence of bias.

Sample data thus allow for testing subjective assumptions. Subjective assumptions can be used to select independent variables and new procedures for computing estimates. By using sample data as described above, those assumptions can be systematically tested.

---

[8] For sets containing only one variable the multiple correlation will be a simple correlation.

The Census Bureau has made some limited use of sample data to assess the accuracy of its postcensal estimates (e.g., Bureau of the Census, 1978), and we encourage the expanded use of this approach. Low-precision as well as high-precision sample data can be used (see Appendix H). The present discussion has been limited in part by lack of development of methodology; more research is needed to extend and refine the approach. More research is also needed to account for errors in benchmarks used for evaluation, whether these benchmarks are sample survey estimates, census estimates, or other kinds of estimates.

3.3 ERROR MODELS

Error in postcensal estimates of population and income can arise from numerous sources. Identifying the different components of error can be valuable for determining where improvement of data or methodology is most needed. In practice, estimates of biases and variances of some error components may be readily available, while only approximate bounds are obtainable for the moments of the remainder of the components. Models of error allow one to combine these different pieces of information to produce estimates (possibly, interval estimates) of the total error. Furthermore, construction of error models leads to insight into ways to improve estimation procedures.

We focus attention here on error decompositions for population estimates provided by linear models, as in the ratio-correlation method or the regression-sample method (Ericksen, 1974; Fay, 1979; Gonzalez and Hoza, 1978).[9] In Appendix G, alternative error models are presented for ratio-correlation estimates. Error models can be constructed in diverse ways, and the particular structure should be chosen to conform both to knowledge about components of error and to desired insights. For example, Appendix G uses error models to analyze the effects of undercount on the postcensal estimates of population obtained under several methods.

We begin by making a simplifying assumption about the forms of the variables in the linear models. The ratio-correlation method uses variables V in the form

$$\frac{V_i(t)/V_i(0)}{V_+(t)/V_+(0),}\ ,$$

where t refers to the current period, 0 refers to the previous censal period,

[9] The following discussion is technical and is mainly for readers familiar with regression theory.

unit i is the area of interest, and a subscript plus sign in place of the subscript i denotes the sum over all areas. The regression-sample method uses variables in the form $V_i(t)/V_i(0)$. For clarity of presentation we will simplify and assume that both methods use variables directly rather than in ratio form.

Both the ratio-correlation method and the regression-sample method use linear models to provide estimates. Error in these estimates arises partly from error in the model used. To make this idea precise, some notation will be introduced. We restrict attention to population estimates. For time t we define

> Y_t vector of n actual (true, unknown) populations;
> X_t matrix $(n \times k)$ of k symptomatic data variables,
>   for the n populations;
> W_t matrix $(n \times n)$ of weights;
> β_t vector of k regression coefficients.

If the optimization criterion is minimization of the weighted sum of squared deviations from fitting Y_t by $X_t\beta_t$, then β_t is given by

$$\beta_t = (X_t' W_t X_t)^{-1} X_t' W_t Y_t, \tag{3.8}$$

where we assume that X_t has rank k, W_t is non-singular, and prime ($'$) denotes matrix transpose. In most applications (including those of the Census Bureau) the matrix W_t is chosen to be the identity matrix, so the weights are all equal. Unequal weighting could correspond to an optimization criterion that placed unequal emphasis on accuracy of the estimates for different areas.

Whatever the optimization criterion, β_t is chosen so that $X_t\beta_t$ provides the best linear fit to Y_t. The difference between the true population Y_t and the best fit $X_t\beta_t$ is the error in the model. Writing the error in the postcensal estimate as $Y_t -$ estimate, we note that

$$Y_t - \text{estimate} = (Y_t - X_t\beta_t) + (X_t\beta_t - \text{estimate})$$

decomposes the error into the error in the model plus another component. The nature of this other component depends on the type of method being used. We first consider this component for the regression-sample method and then for the ratio-correlation method.

For a postcensal time t the regression-sample method uses CPS data for m PSU's to estimate β_t for counties and states. The symptomatic data for the k symptomatic data variables for the m PSU's may be represented by an $(m \times k)$ matrix S, and the sample population estimates for the m PSU's by a vector $\hat{Y}_t$. By using a least-squares optimization criterion, the vector $\hat{\beta}_t$ of fitted regression coefficients may be written as

$$\hat{\beta}_t = (S_t{'}\tilde{W}_t S_t)^{-1} S_t{'}\tilde{W}_t \hat{Y}_t,$$

where $\tilde{W}_t$ is a matrix ($m \times m$) of weights (and we assume that $\tilde{W}_t$ is nonsingular and S_t has rank k). The matrix $\tilde{W}_t$ can be chosen to reflect the different variances of the CPS sample population estimates. Ericksen (1974) sets the diagonal elements of $\tilde{W}_t$ approximately inversely proportional to the sampling variances (i.e., roughly proportional to population size).

The regression-sample method estimates Y_t by $X_t \hat{\beta}_t$. The difference between the best linear fit $X_t \beta_t$ and the fit $X_t \hat{\beta}_t$ estimated with CPS data is

$$X_t \beta_t - X_t \hat{\beta}_t. \tag{3.9}$$

We call (3.9) the error due to data error. This error component can be further decomposed into four data components: error from bias in the CPS estimates, error from random variation in the CPS estimates, error due to differences between the characteristics of the PSU's and the units of analysis (states and counties), and error due to the wrong choice of weights in $\tilde{W}_t$.

The ratio-correlation method uses the same regression model to make postcensal estimates for all t until the next census. These estimates have the form $X_t \tilde{\beta}_0$, where $\tilde{\beta}_0$, the estimated vector of k regression coefficients, is determined on the basis of data from the time $t = 0$ (the previous census year) and X_t is as defined above. If we denote by $\tilde{Y}_0$ the vector of census counts for the n populations and if the optimization criterion is weighted least squares, then $\tilde{\beta}_0$ is given by

$$\tilde{\beta}_0 = (X_0{'}\tilde{W}_0 X_0)^{-1} X_0{'}\tilde{W}_0 \tilde{Y}_0,$$

where $\tilde{W}_0$ is a matrix ($n \times n$) of weights (and we assume that $\tilde{W}_0$ is nonsingular and X_0 has rank k). The weights $\tilde{W}_0$ correspond more closely to W_t than to $\tilde{W}_t$, since there is no sampling variance in $\tilde{Y}_0$ to be adjusted for. Generally, as with W_t in (3.8), the matrix $\tilde{W}_0$ is chosen to be the identity, and thus unweighted regression is used.

The quantity

$$X_t \beta_t - X_t \tilde{\beta}_0 \tag{3.10}$$

measures the difference between the predictions under the best linear fit

$X_t\beta_t$ and the fit obtained with the model derived from the old census data $\tilde{Y}_0$. Although the census estimates $\tilde{Y}_0$ differ from the true values Y_0, in the present context such error is of secondary consideration, and we call (3.10) the error due to structural changes in the regression.

To obtain a decomposition of error, we note that for the regression-sample estimates, the error $Y_t - X_t\hat{\beta}_t$ equals

$$(Y_t - X_t\beta_t) + (X_t\beta_t - X_t\hat{\beta}_t),$$

the sum of error due to the model and of error due to data, where the latter term decomposes into four components (as is discussed above). For the ratio-correlation estimates, the error $Y_t - \tilde{Y}_t$ equals

$$(Y_t - X_t\beta_t) + (X_t\beta_t - X_t\tilde{\beta}_0),$$

the sum of error due to the model and of error due to structural change in regression.

The individual error components can be isolated, and their properties (mean, variance, etc.) estimated. Fay (1979) uses interview data for the Survey on Income and Education (SIE) to study separately the error in the model for regression-based estimates as well as the error from bias and the error from random variation of the SIE estimates of the number of children in poverty. Fay also considers the error component arising from structural changes in regression for the problem of postcensal population estimation. He computes the principal components of various symptomatic indicators of population (school enrollment, labor force, and tax returns) and compares them in different years. This permits analysis of whether prediction models change over time because of changes in the interrelationships of the symptomatic variables apart from changes in their relationship to the dependent variable (population).

To summarize, the error components in linear models appear to be estimable, although much work remains to be done to develop additional theory and methods. We think that further development of methods and appropriate data collection, where practicable, will greatly enhance the ability of the Census Bureau to produce more accurate estimates and to understand the structure of errors. The Panel encourages the Census Bureau to undertake such efforts.

4 Testing Estimates Against the 1980 Census

4.1 POPULATION

The 1980 census presents the first opportunity to extensively test the postcensal estimation methods used in the 1970s. While there are problems with using a decennial census to evaluate estimates (see section 3.1 and Appendix I), such tests are still the most powerful tool for evaluating the accuracy of estimation methods. The first step in performing tests is to state and justify clearly the evaluation criteria to be used (see sections 1.1d and 3.1).

Several basic questions are of interest:

1. How accurate are the estimates of total population and per capita income for different geographic levels (states, countries, subcounty areas)?

2. How does accuracy vary with characteristics of an area, such as population size, rate of population growth, and age distribution of the area's population?

3. How does accuracy vary with time?

4. What other characteristics are associated with the accuracy of the population estimates?

5. Are the estimates biased for certain classes of units?

6. How do the current methods compare in accuracy?

7. How do alternative methods compare? What effects on accuracy would result from modifications to the methods?

8. How much of the error in estimates is attributable to poor data rather than to poor models (i.e., assumptions that do not consistently apply)?

9. How accurate are the estimates of postcensal change in population and in per capita income for different geographic levels? How does the accuracy vary by population size and rate of change in population or per capita income and by other characteristics of the area under consideration?

4.1a CENSUS BUREAU PLANS

The Census Bureau has prepared an outline of possible tests of the methods against the 1980 census.[1] The outline is extensive, and a partial summary follows. Questions 1, 2, 5, and 6 above are considered through summary tabulations similar in form to Tables 2.1–2.11. Other tabulations can also be done to study separately the accuracy of the estimates for areas undergoing boundary changes and annexations. Question 3 is considered by focusing on areas that received special censuses during the 1970s and comparing the deviations of the estimates from the special censuses with the deviations of the estimates for April 1, 1980, from the 1980 census counts.

A variety of alternative methods are mentioned in the outline as candidates for testing; these include methods described in the literature but not now used by the Bureau, as well as new methodology being developed at the Bureau. In particular, at the subcounty level the housing unit method may be tested for the approximately 16,000 subcounty areas for which the requisite data are available.

To consider question 8, the Bureau may recompute estimates by using (where possible) census data in place of administrative data and comparing deviations of these estimates with the deviations of the original estimates from the 1980 census.

A variety of possible modifications to methods are presented in the outline. For county estimates these include use of optimal weights and alternate ways of performing the "rake/float" adjustments (described in Appendix A, section 4.2). For the AR method, possible modifications include adjusting the migration rate computations for differential filing patterns by race and expanding the time intervals between matched years. For the subcounty estimates the possible modifications also include computation of the AR estimates separately for populations under

---

[1] This outline is T80-CV, *Explanatory Notes on "1980 Tests—Outline"* and *1980 Tests—Outline,* an unpublished working document of the Census Bureau.

65 and 65 and over and elimination of county controls. For subcounty estimates for places with population of less than 1,000, possible modifications of the AR method include using the migration rate for a larger area to estimate a given area's migration rate, estimating the percent change in population to be proportional to the percent change in the number of total exemptions claimed on federal individual income tax returns, and assuming zero migration.

4.1b PANEL COMMENT

The outline of tests is ambitious, and it is unlikely that resources will allow the Census Bureau to conduct all the tests. This section presents the Panel's comments and suggestions for the tests.

The decomposition of total error into data error and error in the model (see question 8), as considered by the Bureau in the outline, is important. The prime candidate for such a decomposition is perhaps component method II. The likely approach would be to first estimate total error from the deviations of the 1980 CM II estimates from the census results. Alternative estimates could also be prepared with the same methodology but using census data instead of symptomatic data. Under certain assumptions, error in the alternative estimates—estimable from the deviations of these estimates from the 1980 census results—may be thought of as error in the model. The difference between the total error and the error in the model is the data error. Such a decomposition will be helpful in determining how to improve the estimates. Work is needed, however, to try to estimate the effect of error in the 1970 and 1980 census data. In the description just presented, this error is incorporated in the component of error attributed to the model.

A similar decomposition of error should be performed for the ratio-correlation method. The total error in the RC estimates may be decomposed into the error in the model and the error due to structural change in the regression (as described in section 3.3). Total error may be estimated from the deviations of the 1980 RC estimates from the census results. An alternative regression equation can also be constructed from symptomatic data and census data for 1970 and 1980. The deviations of the estimates yielded by this alternative equation from the 1980 census results form the basis for estimating the error in the model. The difference between the total error and the error in the RC model is the error due to structural change.

Optimal weights for the averaging of estimates can improve the accuracy of estimates over that of equally weighted averages. (A method of determining optimal weights is discussed below.) Optimal weights for averaging should be computed, and the accuracy of optimally weighted averages should be compared with the accuracy of equally weighted averages.

Modification to the AR estimates of migration rates to adjust for differential filing patterns by race (which are discussed in section 5.1b (1)) should be tested. Other proposed modifications should also be tested, including one to estimate the percentage change in population to be proportional to the change in the number of exemptions claimed by individuals on IRS forms.

The Census Bureau should also test whether the AR method performs better when the time intervals between matched years are longer. Simple models (developed in Appendix G) suggest that the method performs better when the intervals are shorter, but those models may be too simple. In performing such tests on the AR method it may be cost effective to use samples of the tax files. Rather than randomly sampling tax returns, it may be better to select a sample of areas and then create a subfile of data for the sampled areas. For areas with large populations, samples rather than complete sets of tax return data could be used; for selected small areas the complete set of data would be used.

It is extremely important to understand how accuracy varies with time, and the Bureau should conduct the analysis of error rate trends described in its outline. The Bureau's proposed analysis would be restricted to counties and places with special censuses conducted some time in the 1970s. The approach would be to compare the deviations of estimates during the 1970s from the special censuses with the deviations of the 1980 estimates from the 1980 census. It is important for such an analysis that the 1980 estimates be computed such that the previous special census data are ignored. The analysis may be problematic because of the nonrandom sample of areas and time points, especially with regard to boundary changes and annexations.

It is also important to evaluate the accuracy of estimates of postcensal change in population and per capita income. The relative errors for estimates of postcensal change are much larger than those for estimates of total population or per capita income. The patterns of error are strikingly different, and it is postcensal changes in population and per capita income that produce changes in general revenue sharing allocations.

Test results should be analyzed for areas classified and cross-classified by size, percent change in population, and geographic level. Modern techniques of data analysis should be used to identify additional variables that are associated with the estimates. Summary tabulations (e.g., Tables 2.1–2.11) are a useful way to present the test results, but the use of models to study the associations between accuracy and area-population characteristics is highly recommended.

4.2 PER CAPITA INCOME

Tests of the Census Bureau's per capita income (PCI) estimates for states, counties, and subcountry areas should be conducted in conjunction with the 1980 census. These tests should not only examine the Bureau's money income concepts but also determine the adequacy of the personal income series of the Bureau of Economic Analysis (BEA) for counties and states.

A test should be conducted for all counties and for a probability sample of subcounty areas with tabulations classified by population size of area, rate of change in population, and, for counties, proportion of population that is rural (test 1, in Table 4.1). A second test should determine the compatibility between the Census Bureau's postcensal estimates as now computed and the personal income estimates compiled by BEA. The BEA data will have to be adjusted to make the income base compatible with the Census Bureau's money income estimates (test 2, Table 4.1). A third test would involve comparing the 1979 BEA adjusted data with the 1979 census PCI. Tests should be conducted for large

TABLE 4.1 Proposed Tests for PCI Estimates

Test	Data Source	1979 PCI Estimate
1	1979 postcensal PCI	1979 census PCI
2	1979 postcensal PCI	BEA_a*/1980 population
3	$\dfrac{1979\ \text{BEA}_a}{1980\ \text{population}}; \dfrac{1979\ \text{BEA}_u{}^*}{1980\ \text{population}}$	1979 census PCI
4	$\dfrac{1979\ \text{BEA}_a}{1969\ \text{BEA}_a} \times 1969\ \text{census PCI}$	1979 census PCI
5	$\dfrac{1979\ \text{BEA}_u}{1969\ \text{BEA}_u} \times 1969\ \text{census PCI}$	1979 census PCI
6	regression PCI estimates	1979 census PCI
7	regression BEA_u/POP	1979 postcensal PCI

*Adjusted to approximate total money income; uses BEA_a for adjusted, BEA_u for unadjusted.

counties and small counties, for counties with an insignificant amount of farm income and counties for which farm income is very important, and for stable and rapidly changing counties.

Other tests should be designed to determine if BEA adjusted data can be used to generate relatively simple census PCI measures for counties. These tests should be conducted by determining the growth rate in each county's BEA personal income figure over the period 1969-1979. Once this growth factor has been determined, it could be applied to the Census Bureau's PCI for 1969. By multiplying the 1969 PCI against the growth factor, an effort could be made to determine how closely this coincides with the census PCI for 1979 (test 4, Table 4.1). Again, this test should be designed for counties with a large proportion of farm income and for counties for which far income is insignificant. Counties with a large population should be included in the tests and compared with counties with smaller populations. The accuracy of estimates for rapidly changing counties should be compared with the accuracy for more stable counties.

The same tests could be repeated by using unadjusted BEA data and by comparing it to the 1980 census estimates of 1979 PCI (tests 3 and 5, Table 4.1). While there are reasons for adjusting BEA data, the imputed rents and in-kind income may not be as significant as some believe. Tests should be done to determine if the unadjusted data produce as reasonable an estimator as do the adjusted data.

Other tests should be conducted to determine if it is feasible to use regression techniques for measuring income changes. Per capita income could be regressed against such factors as the proportion of total income in wages and salaries, the proportion of total income in farm income, the age structure of the population, employment in the county, tax returns filed, and other economic variables (test 6, Table 4.1). It might be feasible to construct an income estimating program similar to the ratio-correlation estimating program for population for state and county units. Regression techniques might also be used for generating postcensal estimates of per capita personal income for counties (test 7, Table 4.1). The regression estimates might be able to be produced more quickly than the usual personal income estimates. (It is doubtful that the regression technique could be extended to subcounty units.) These tests should be conducted against both the 1979 census income estimates and the postcensal 1979 PCI estimates.

5 Technical Critique of Methods of Estimation

5.1 CRITIQUE OF INDIVIDUAL METHODS OF ESTIMATION

In general, the individual methods used by the Census Bureau to estimate population—component method II (CM II), ratio-correlation method (RC), administrative records method (AR)—appear sound. The method used to estimate per capita income is also sound. As was discussed earlier, however, the methods can produce highly inaccurate estimates for small areas, largely because of the lack of adequate data for producing accurate estimates. The Panel does not know of any methods—short of prohibitively expensive sample surveys, annual censuses, or population register systems—that would yield significantly more accurate estimates on an annual basis.

This chapter presents the Panel's specific criticisms of the three methods for population estimation (CM II, RC, and AR) and also of the method for income estimation. When it is possible, alternative procedures and tests are suggested. (The reader is referred to Appendix A for details on the population estimation methods and to Appendix B for a summary of the income estimation method. The reader may wish to refer to those appendices in following the details of the criticisms presented here.) While our suggestions cannot solve the problem of inaccurate estimates for very small areas, they would lead to qualitative improvements that may prove to be significant for some areas.

The Panel recognizes that many of our criticisms are minor and pertain (for the population estimates) to logical and methodological irregularities in Census Bureau procedures. The Panel was unable to discover methodological solutions for the most serious problems in making good estimates for the very small areas, especially the problems of attributing the correct geography to the individual income tax returns used by the AR method, adjusting the estimates for boundary changes and annexations, and estimating births and deaths for those areas for which data are not available.

5.1a COMPONENT METHOD II

Our critique of CM II concerns the procedures for estimating net migration. In particular, our discussion focuses on proper denominators for migration rates. The component method II estimates net migration from data on changes in school enrollments. We have two technical criticisms of this approach: tests are needed to determine whether the criticisms are quantitatively significant.

The migration "rates" used in CM II for state and county estimates are not consistent. The denominators for the school-age and female migration rates for the period preceding the census (the precensal rates NGQMIGYRAT(0) and SCLMIGRAT(0) discussed in Appendix A, sections 2.2 and 3.4d) consist of the relevant population in the particular location enumerated in the census (that is, at the end of the precensal migration interval). Use of such a denominator need not necessarily cause any problem. But the denominator of the postcensal school-age migration rate is the expected survivors (assuming births and deaths but no migration) of school age in the particular location at the midpoint of the postcensal interval. Hence the denominators of the two measures do not conform. The denominators would be consistent if the actual school-age population at the end of the postcensal period, not the expected survivors (assuming natural increase but no migration) in the middle of the period, were employed.

The postcensal migration rate R would then properly be a rate whose numerator is the number of surviving net migrants and whose denominator is the number of inhabitants of the area at the end of the period. Hence if R were multiplied by an appropriate base, the result would be the number of surviving migrants at the end of the period. But what is needed for the migration component of CM II is the total net number of migrants during the interval, regardless of whether the inmigrants or outmigrants survived, since the methodology assumes that

$$P_1 = P_0 + B - D + I - O,$$

where all events are charged against that area and where P_1 is post-censal population, P_0 is censal population, and B, D, I, and O refer to the total numbers of births, deaths, inmigrations, and outmigrations, respectively, occurring in the area during the period.[1] There does not appear to be an easy procedure for converting a migration rate pertaining to surviving net migrants into one measuring total net migrants. Although they are certainly not equivalent, an easy solution would be to assume that they were. In that case the proper base to multiply R by in order to estimate total net migration would be

$$\text{BASE} = P_0 + \tfrac{1}{2}(B - D) + \tfrac{1}{2}M,$$

where $M = I - O$. Then since M equals R times BASE,

$$\text{BASE} = [P_0 + \tfrac{1}{2}(B - D)]/(1 - R/2).$$

The difference between the estimates of migration using BASE as just defined[2] and the base population currently used by the Census Bureau, $P_0 + \tfrac{1}{2}(B - D)$, is

$$\text{BASE} \cdot R^2/2. \tag{5.1}$$

This difference is obviously nonnegative, regardless of the sign of R. Because the estimates of county (state) population are controlled to state (national) totals, the final effect of using BASE as a base population figure is not obvious. We suspect the effects will generally be minor, but in some cases they will not be negligible, especially when there is wide variation in R. A simple numerical example will illustrate this. Suppose one is estimating the population of 50 areas (e.g., states or counties in a state) and suppose that all areas have the same population P_0 at the beginning of the estimation interval and also that there is a constant proportional natural increase, S (less than 10 percent).[3] Consider three classes of areas for which the values of R are 0.15, 0.10, and 0.01, respectively, and let the numbers of areas in these classes be 3, 4, and 43. (This distribution of R over areas approximates the distribution of the 5-year migration rates for 1970–1975 for states, as published by the Bureau of the Census (1976, Table 1) except that here we always take

---

[1] Group quarters populations are ignored in this discussion.
[2] Here we refer to the estimates before higher-level controls are applied.
[3] This last assumption has minor impact but simplifies the analysis; larger values and nonconstancy of S would not alter implications substantively.

R positive.) The increase in the uncontrolled population estimates for each area, by class (from (5.1)) would be

$$
\begin{array}{ll}
0.01125 \times \text{BASE} & R = 0.15 \\
0.005 \times \text{BASE} & R = 0.10 \\
0.00005 \times \text{BASE} & R = 0.01
\end{array}
\qquad (5.2)
$$

when we use the proposed BASE method instead of the base population used by the Census Bureau. Substituting $P_0/(1 - R/2)$ for BASE , (5.2) becomes

$$
\begin{array}{ll}
0.012162 P_0 & R = 0.15 \\
0.005263 P_0 & R = 0.10 \\
0.00005 P_0 & R = 0.01.
\end{array}
\qquad (5.3)
$$

Thus the increase in the sum of the area populations is $P_0 [3(0.012162) + 4(0.005263) + 43(0.00005)]$, or $0.0597 P_0$. The effect of controlling is to reduce the uncontrolled estimate for an area by $0.0597 P_0 f$, where f is the ratio of the uncontrolled estimate for the area to the total of all uncontrolled estimates. Since the uncontrolled estimate for an area is given by $P_0(1 + S) + RP_0/(1 - R/2)$, the respective values of f for the three classes of areas are 0.023, 0.022, and 0.020.[4] When $0.0597 P_0 f$ is subtracted from the figures in (5.3), the final effect (i.e., after controls) of using BASE rather than the base population used by the Census Bureau is to change the estimates by

$$
\begin{array}{ll}
0.011 P_0 & R = 0.15 \\
0.004 P_0 & R = 0.10 \\
-0.001 P_0 & R = 0.01.
\end{array}
\qquad (5.4)
$$

Note that use of the proposed BASE decreases slightly the population estimates for areas with R closest to zero. More important is the not negligible increase (1.1 percent of P_0) in the population estimates for areas in which R is largest ($R = 0.15$). Although the values of R in this example were all positive, similar results would hold for negative values of R; e.g., if R were -0.15 instead of $+0.15$ for the three areas in our example, use of BASE would still increase the estimate by about 1 percent of P_0.

---

[4] The values increase as S increases: from 0.0226, 0.0215, and 0.0197 for $S = 0.0$; to 0.0229, 0.0219, and 0.0202 for $S = 0.1$.

5.1b ADMINISTRATIVE RECORDS METHOD

The administrative records method (AR) is a component method of estimation that develops migration estimates on the basis of the numbers of
tax returns matched across years according to social security numbers.
(The method is described in detail in Appendix A.) Our principal criticism of the AR method concerns estimation of net inmigration. The
administrative records method develops migration estimates on the basis
of changes in address on tax returns, and there is the possibility of
biases arising from different filing rates for different segments of the
population. The bias in the estimate may be severe for some areas, but
there is insufficient evidence to draw a conclusion. Our other criticisms
are qualitatively minor. We suspect they are not quantitatively significant, but only tests can determine if this is true.

5.1b(1) *Biases*

To calculate the net number of migrants to an area, the AR method
multiplies the estimated migration rate by a base population figure. The
net inmigration rate (IRSRAT) is estimated by

$$\text{IRSRAT} = \frac{\text{INS} - \text{OUTS}}{\text{OUTS} + \text{NONMOV}},$$

where INS, OUTS, and NONMOV refer to numbers of exemptions on
IRS tax returns matched by social security number across 2 years (see
Appendix A, section 3.7 for precise definitions). This calculation ex-
cludes those segments of the population not represented by exemptions
on matched tax returns. Because the excluded populations often have
different migration patterns than those covered by the tax returns, the
estimates of net migration can be biased for many areas. Excluded
populations tend to include, disproportionately, many aged, minority,
and, possibly, low-income people.[5] In a recent report (Bureau of the Census, 1978), David Word outlined a method for adjusting the migration
estimates to remove the biases. The technique is summarized below; in
the Panel's view, the technique shows promise and should be evaluated
when 1980 census results are available.

This technique defines the coverage ratio as the ratio of exemptions

[5] Exclusion of persons over 65 is not significant for AR estimates at the state and county
levels, where Medicare data rather than tax data are the basis for the estimates.

contained on tax returns filed by persons under 65 years of age to civilian population under 65. The 1970 national coverage ratio for blacks was 82.2 percent, and that for whites a surprising 101 percent (Bureau of the Census, 1978). This anomaly for whites is explained by the claiming of multiple exemptions for individuals and by undercount in the 1970 census.[6] It does not imply that the entire white population was covered by the tax system. Also important in this method is the match ratio, defined as the ratio of the number of tax returns matched for the 2 years to the average of the number of returns filed in both years. Following Bureau of the Census (1978), we define the efficiency ratio as the product of the match ratio and the coverage ratio: efficiency ratio = match ratio × coverage ratio. The efficiency ratio roughly indicates the fraction of the population for which we have data to estimate internal migration.

Efficiency ratios vary significantly by age, race, sex, and income. Error is introduced into IRSRAT because area migration rates also vary by age, race, sex, and income. For example, Word estimated the efficiency ratios in Mississippi for 1970-1975 at 85.8 percent for whites and 42.7 percent for blacks. Assuming that the efficiency ratio for each race did not vary by migration group (inmigrants, outmigrants, nonmovers), Word calculated that the net migration for Mississippi for 1970-1975 was estimated too high by 22,000 persons by the AR method—roughly 1 percent of Mississippi's 1970 population.

It is suspected by some that for large cities the estimates are too low: migrants into central cities tend to be young, nonwhite, and low-income persons, and many of the inmigrants to the cities file tax returns for the first time after they migrate (see Mann, 1978). If this is true, then the efficiency ratio for inmigrants would tend to be lower than the ratios for nonmovers or outmigrants from the large cities, so that net migration to large cities is underestimated. Lowe et al. (1974) compared the AR estimates to special censuses taken in Washington and found that the AR method tended to overestimate the populations of surburban places (municipalities within 30 miles of metropolitan cities). They also found that the AR method underestimated the population of cities and towns with large proportions of agricultural, construction, or mining workers.

Full evaluation of the biases in the AR estimates must await the results of tests against the 1980 census. If these test results support the reasoning above, then adjustments for bias should be considered. One possibility to stratify tax returns by variables i (e.g., race, age, income, etc.) and calculate INS, OUTS, and NONMOV separately for each stratum.

---

[6] For example, some persons in high school or college file tax returns to obtain refunds and are also counted as dependents on their parents' tax return.

Then the net migration rate would be estimated by

$$\frac{\Sigma(\text{INS}_i - \text{OUTS}_i)F_i}{\Sigma(\text{OUTS}_i + \text{NONMOV}_i)F_i},$$

where F is the reciprocal of the efficiency ratio.

The Panel recognizes that there are practical problems to stratifying by variables i. For example, the tax returns themselves do not provide information on race or age (other than over 65 or under 18). In the study described by Bureau of the Census (1978), race information was obtained by matching the IRS file with a sample of the Social Security Administration's summary earnings file. The latter file can also be used to provide age data. We commend the Census Bureau for their efforts to develop ways to adjust for biases in the AR method, and we encourage further work to extend the techniques to counties and cities with moderate to large populations. The usefulness of adjustment techniques such as the one described above should be determined by tests of adjusted estimates against 1980 census results.

5.1b(2) *Central Rates*

The denominator of the migration rate calculated on the basis of the IRS returns is the number of nonmovers plus the number who moved away during the time interval under consideration. The rate is applied to a base population defined as the initial population plus half the births minus half the deaths and minus half the net number of immigrants from foreign countries.[7] Both the numerator and the denominator of the rate are derived from the number of exemptions in the *final* year of the time interval. Hence the rate is not a central rate (m_x). To convert the migration rate into an approximate central rate, several changes are needed:

1. The INS, OUTS, and NONMOV should be measured as the average of number of exemptions listed in both returns.

2. The denominator would then consist of NONMOV and half the sum of INS and OUTS, where all are defined as in point 1.

3. The old base should be divided by a factor $1 - \text{IRSRAT}/2$ to yield a new base that equals the old base plus half the net migration during the period. This base should not include half those aliens who immigrate

---

[7] For state and county estimates the base population also includes half the net movement from military barracks to non-group quarters.

(net) into the region, since immigrants are specifically counted separately. The initial population should exclude those living in group quarters.

4. Both the denominator of the rate and the base would then be based on a person-years of exposure concept.

5. If points 2 and 3 but not 1 are applied, then the estimate of migration would be identical to that of the Census Bureau.[8] Thus simply instituting the change in point 1 and following the census procedure thereafter will produce a result equivalent to points 1, 2, and 3.

5.1b(3) *Migration Controls*

In the component methods the final estimate of migration is a residual. The initial population estimates are controlled to the population total of the next higher unit (subcounties controlled to counties, counties to states, states to national). Births and deaths are accepted as being completely registered (or estimated at the subcounty level), and the final net migration component is simply the difference between the final estimate of population and the initial population estimate updated by births and deaths. It is clear, however, that if the initial population estimate, births, and deaths are accepted as being correct, then it logically would make more sense to control the migration component itself (rather than population total). An alternative procedure exists for such a control when separate estimates of inmigrants and outmigrants are available. Hence the following discussion applies only to AR, and not to component method II or ratio-correlation method.

Let MIGIN be uncontrolled estimate of inmigrants from IRS, and let MIGOUT be uncontrolled estimate of outmigrants from IRS; thus MIGIN = [INS/(OUTS + NONMOV)] × MIGBASE. If ΣMIGIN − ΣMIGOUT is supposed to equal a control value K but does not, then one simply finds

---

[8] The Census Bureau's current procedure estimates net internal migration by

$$M = \frac{\text{INS} - \text{OUTS}}{\text{NONMOV} + \text{OUTS}} \cdot [\text{POP}(0) + \tfrac{1}{2}(B - D) + \tfrac{1}{2}\text{IMMIG}].$$

If points 2 and 3 but not 1 are applied, the net internal migration estimate is

$$M^* = \frac{\text{INS} - \text{OUTS}}{\text{NONMOV} + \tfrac{1}{2}(\text{INS} + \text{OUTS})} \cdot \frac{[\text{POP}(0) + \tfrac{1}{2}(B - D)]}{1 - \tfrac{1}{2}\left[\dfrac{\text{INS} - \text{OUTS}}{\text{NONMOV} + \tfrac{1}{2}(\text{INS} + \text{OUTS})}\right]};$$

Hence $M = M^*$.

some proportion a such that $(1 - a) \cdot \Sigma\text{MIGIN} - (1 + a) \cdot \Sigma\text{MIGOUT} = K$. This yields

$$a = \frac{\Sigma\text{MIGIN} - \Sigma\text{MIGOUT} - K}{\Sigma\text{MIGIN} + \Sigma\text{MIGOUT}}. \tag{5.5}$$

The controlled or final estimates of inmigration and outmigration for each area would become $(1 - a)\text{MIGIN}$ and $(1 + a)\text{MIGOUT}$, respectively. Note that a will normally be very small, since net migration is normally small in relation to gross migration. As an example of the use of (5.5), consider the migration components for states: they must sum to zero, since internal migrants from one state must go to another. If $\Sigma\text{MIGIN} - \Sigma\text{MIGOUT}$ does not equal zero, a small adjustment would force this total. For subcounty (county) areas, ΣMIGIN and ΣMIGOUT are adjusted to sum to the estimated migration component for the county (state). Thus in AR, births, deaths, and all alien immigration would be accepted as if they were true, and only the group and non-group migration would be scaled to sum to a given total yielded from the final estimate for the next higher geographic level.

This proposed method of controlling may produce nontrivial changes in estimates for areas in which the current method for controlling to totals produces changes in the area's total population that are large in relation to the estimated net internal migration.

5.1c RATIO-CORRELATION METHOD

The ratio-correlation method is widely used, and its application by the Census Bureau suffers from the same problems found elsewhere. The procedure assumes that the vector of regression coefficients for symptomatic variables is invariant from the immediately preceding intercensal period to the postcensal period in question. However, this invariance does not hold over time, both because of structural changes in the underlying relationships of the variables and because the quality of the symptomatic data varies.

The problem of structural changes in the variables' relationships is widely appreciated (e.g., Bureau of the Census, 1974). This problem may be most severe when the number of variables in the ratio-correlation equation is large, so that collinearity becomes important. For example, Namboodiri and Lalu (1971), in a test of ratio-correlation estimators for counties in North Carolina, found that the average of five univariate regressions produced more accurate estimates than did the fitted five-variable equation. The explanation is that although the five-variable equation is best in the base period, this optimality need not hold over time, since the relationships of the symptomatic variables to each other and to the variables of interest change. To resolve this problem, more research is needed. Use of current sample data, as discussed in section 3.2, is one approach. Additional insight may be provided by multivariate techniques such as principal components analysis (see Fay, 1979, pp. 179–183).

5.1d ISSUES IN INTERNATIONAL MIGRATION

The estimated total net number of international migrants is distributed first among states and then among places in states. The estimated net national number of immigrants is obtained by taking the number of legal immigrants reported by the Immigration and Naturalization Service and subtracting a constant of 36,000 as the number of emigrants.[9] For example, the resulting net number of international migrants was 343,000 for calendar year 1978 (Bureau of the Census, 1979). The allocation among states is determined by the intended residence claimed by legal immigrants on forms collected by the Service. Place of residence is also coded if its population exceeds 100,000. Allocation among places proceeds in two steps. From the forms, the fraction of immigrants intending to reside in places of over 100,000 is determined. This fraction times the estimated net number of immigrants is allocated among places exceeding 100,000 in the same proportions given by the forms. The remaining net number of estimated immigrants is allocated among places not exceeding 100,000 on the basis of the distribution of the foreign born recorded in the 1970 census. County estimates are obtained by summing estimates of places within counties.

This description identifies several sources of error, which in certain instances (discussed below) might cause serious distortion in the population estimates. First, to the extent that the place of intended residence is not the same as the place of actual residence, there will be distortions in the allocation of the national net number of international migrants. Further, the allocations are based on information obtained from immigrants; undoubtedly, the geographic distribution of emigrants is different.

Second, changes in the total net number of immigrants would result in the same relative net allocation of immigrants to places but would

[9] The Immigration and Naturalization Service stopped collecting data on alien immigration in 1957; permanent departures of U.S. citizens are also not recorded (see section 1.2c of Appendix A for more discussion).

alter the relative distribution of total population, since net immigrants are not a constant fraction of total population across the country. Hence changes in the constant used to estimate the number of emigrants would affect, although to a small extent, the estimated distribution of population.

Third, net illegal immigration is ignored entirely in the estimate of net international migration. If some number were assumed and added to the total net figure, the allocation of net immigrants among places would be increased proportionally, but the distribution of population would change (as above). However, illegal net immigrants almost certainly do not have the same settlement patterns as those given by the forms obtained from legal immigrants. Hence allocations among states are almost certainly distorted. The misallocation within states is mitigated to the extent that illegal immigrants were counted as being foreign born in the 1970 census and to the extent that settlement patterns have not changed over the decade. However, even such a qualitative assessment is risky because of the two-stage allocation procedure and because there are few data to provide the basis for judgement. In addition, to the extent that illegal immigrants were not counted in the 1970 census, the base for the updates is relatively too small. This comment applies to all places that were underenumerated in 1970, regardless of whether those not counted were illegal aliens.

The net result of these considerations is that the population data for states with heavy concentrations of illegal immigrants are subject to downward bias. The problem for individual places, including such large cities as New York, Los Angeles, and Houston, may be severe. These places are thought to attract a large fraction of the illegal immigrants to the United States, yet the estimating procedure has no mechanism to take account of such new immigrants. Since these cities (and others) may be constrained by the 145-percent rule (see Appendix E), the gain in GRS revenue that would occur if immigrants were properly allocated could be significant. It must be noted, however, that in spite of these obvious shortcomings we can think of no better procedures for allocating net immigrants. Until the Immigration and Naturalization Service obtains adequate data on legal emigrants and immigrants and until illegal migration becomes numerically trivial or better techniques are developed to estimate it at the national and local levels, adequate estimates cannot be produced.

5.1e PER CAPITA MONEY INCOME

Income is a complicated concept. Different measures of income can be constructed, depending on the concept adopted. The Census Bureau uses a concept of per capita money income, which represents the mean or average total money income of residents in a given area at a given point in time. Total money income is the sum of six components: wage and salary income; nonfarm self-employment income; farm self-employment income; social security and other retirement income payments; public transfer payments, including assistance payments; and other income sources, including interest, dividends, unemployment insurance, etc. Data on money income can be easily obtained from household surveys such as the Current Population Survey and from decennial censuses. However, many of these data are difficult to obtain from administrative records, which form the data base from which estimates of postcensal changes in income are developed. In updating county per capita income the Census Bureau relies on adjusted estimates of county farm income and other income components (other than wages and salaries) produced by the Bureau of Economic Analysis (BEA). The only available data for subcounty estimates are the adjusted gross income (AGI) figures on the IRS individual income tax returns; the BEA county estimates for the remaining components of income are apportioned to subcounty areas.

The conceptual basis for the BEA income estimates is that of personal income in the national income accounts. The national accounts measure income generated by various kinds of economic activity; personal income is the income of all residents of an area from all sources (including in-kind payments and imputed items). It includes income received not only by individuals but by quasi individuals (nonprofit institutions, private noninsured welfare funds, and private trust funds). The Census Bureau's money income is a statistical construct designed to be susceptible to measurement in household interviews.[10] The two concepts, money income (Census Bureau) and personal income (BEA), are not congruent, and converting BEA personal income estimates into estimates of components of total money income requires difficult adjustments of unknown reliability.

Farm income is especially difficult to estimate. The Census Bureau defines farm self-employment income as the gross income received from operation of a farm minus production expenses. The BEA farm income estimate measures income arising from the current year's production in

---

[10] Neither the BEA's personal income concept nor the Census Bureau's money income concept adequately measures income as a return to a factor of production. The issue is most serious with respect to self employment income. All proprietary establishments require labor and capital inputs, but the final profit or loss figure represents a net sum. If the returns to labor and capital were clearly identified and measured as a capital return or a labor return, an improved income measure would result. This income measure would undoubtedly differ from current accounting income measures.

the farm sector. Thus the BEA gross farm income includes cash receipts from farm marketing of crops and livestock, payments to farmers under the several government support programs, the value of food and fuel produced and consumed on the farms, the gross rental value of farm dwellings, and the value of net change in inventories of all crops and livestock.

The first two items, cash receipts from marketing of crops and livestock and payments to farmers under the several government programs, are the most important components of gross farm income, and they are also included in the money income concept of the Census Bureau. However, with the exception of a few states, annual data on cash receipts for crops are available only at the state level. The available data on cash receipts must therefore be disaggregated for counties. These disaggregations are made by prorating current cash receipts by crop according to the past receipts measured by the last quinquennial census of agriculture. This procedure presents a data problem because 3 or 4 years after an agricultural census a small area such as a county may have shifted its production from one crop to another. A crop that was important at the time of the census of agriculture may be far less important within a short time period because of changing market conditions. An excellent example is soybean production.[11] Less dramatic production shifts, while more subtle, may destroy the accuracy of the process of apportioning a state estimate among counties. The problem is aggravated when farm income is a large proportion of a county's total income.

Inventory adjustments present another difficult issue for the measurement of farm income. The Census Bureau's money income concept does not make allowance for inventory changes. If inventories increase during a given time interval, the Census Bureau's income measure could well be negative, since sales or receipts would be down with expenses constant, while the BEA's measure of net farm income could show a positive total because it makes allowances for inventories. Omission of inventory changes produces erratic shifts in the Census Bureau's estimates of county farm income, a measure notorious for fluctuations. Substantial shifts in farm income (or any type of income) are desirable in the data series when they correspond to real shifts in production. Part of the problem with the Census Bureau's farm income results from a faulty conceptual base.

[11] In many areas of the country, cotton and soybeans can be planted interchangeably. They are close substitutes in production, and production changes occur rapidly in response to price changes. If the price of cotton is high, farmers plant cotton. If the price of soybeans is high, farmers plant soybeans. If there is any one point well established in the literature, it is that farmers are extremely responsive to changes in the relative prices of their crops (Schultz, 1964).

This problem extends beyond updating, for if inventory accumulation or release were abnormal at the time of the decennial census, the base figure for farm self-employment income would be distorted.

Other adjustments in the BEA farm income measure must be made before the farm income component can fit into the money income concept mold. The value of imputed rent on farm dwellings and the value of home consumption of food and fuel must be excluded from the BEA data to fit the Census Bureau's money income measure. These adjustments are easy to perform, however, and can be done satisfactorily.[12] The BEA data must also be attributed to place of residence, but these adjustments are performed anyway (Bureau of Economic Analysis, 1977).

The suitability of a particular concept of income depends upon the uses to which it is put. The general revenue sharing formulas use money income in several ways. In the five-factor formula for states, the product of population and the reciprocal of per capita income serves as a measure of "relative poverty." In the three-factor formula for states and in the county and subcounty formulas, "need" is measured by the inverse of per capita money income. Total money income—estimated as the product of the population and per capita income estimates—is used in the county and subcounty formulas to measure the relative financial ability of a county or subcounty government to collect taxes. In contrast, in both the three-factor and five-factor formulas the relative financial ability of a state government to collect taxes is measured by total personal income. The Panel has not considered which of the two income concepts is more appropriate for each of these uses, except to note that neither the money income concept of the Census Bureau nor the personal income concept of the BEA is ideally suited to represent the variables above. For example, both income measures are only partial indicators of a government's financial ability to collect taxes because they fail to reflect revenue sources, such as motor fuel taxes in tourist-oriented states like Vermont or gambling in Nevada (Advisory Commission on Intergovernmental Relations, 1971). Also, as measures of "need" or "relative poverty," both income measures fail to reflect area differences in the cost of living, the types of services needed by area residents, or the income distribution in an area (Office of Federal Statistical Policy and Standards, 1978, p. 30).

Considering the weak conceptual basis of the Census Bureau's money income measure and the complex and not inexpensive adjustments re-

---

[12] The BEA income estimates involve the addition of these components, which are difficult to measure, but the adjustment for compatibility with Census Bureau income merely involves subtracting these components from the BEA measure.

quired to produce updates of county money income, the Panel recommends that the Census Bureau seriously consider not producing postcensal per capita money income estimates for counties. Alternatively, the Census Bureau could consider simpler procedures for making updates; some possibilities were suggested in section 4.2. If tests against the 1980 census show that the current methods or one of the simpler methods yields highly accurate estimates, then the Census Bureau may wish to continue making postcensal per capita income estimates for counties. Otherwise, users should rely on BEA estimates of per capita personal income by place of residence rather than on per capita money income.

A separate argument for using BEA personal income estimates is their conceptual linkage to the national income and product accounts. Movements in the level of economic activity are monitored through the national income and product accounts system. Policy decisions associated with economic activity rely on such concepts as gross national product, personal income, or one of the other accounts. If it is assumed that the BEA local data are as reliable or more reliable than the money income data, a point that requires testing (and such testing would admittedly be difficult), BEA data have an edge due to their consistency with other recognized measures of economic activity.

Reliance on the BEA estimates is not viewed by the Panel as a panacea. No tests of personal income estimates have been performed, and their accuracy has not been measured. However, the Census Bureau county money income estimates draw heavily on the BEA estimates, so errors in the latter are likely to be present in the former. Estimation of county income, under either concept, is difficult for those components such as farm income, for which good area data are not available. Subcounty postcensal income estimates should not be produced on either a personal income or money income basis, because the concepts and methods are too complex and the communities too numerous to produce reliable income estimates. If a mid-decade census is taken, subcounty income updates could be produced on a quinquennial basis.

5.2 CRITIQUE OF COMBINATIONS OF METHODS: ISSUES OF UNIFORMITY AND AVERAGING

5.2a ALTERNATIVE METHODS OF ESTIMATION

There are three general kinds of procedures that could be used to obtain local estimates: (1) using results from the most recent decennial census (without updating), (2) using sample data to produce direct or synthetic estimates, and (3) using auxiliary data according to a model. The first procedure is used at the state level for determining representation in the electoral college, which changes only when decennial census tabulations are published. The second procedure was used for estimates of variables collected on a sample basis in the decennial census; this procedure is used to obtain the 1970 census estimates of per capita income. The third procedure is used by the Census Bureau to compute postcensal estimates of population and per capita income. Choices among ratio-correlation, administrative records, and component method II estimates pertain only to the selection of the best representative of this category of estimates.

It is also possible to suggest a fourth procedure, which would involve an optimal combination of the second and third (and, possibly, the first) procedures. One way of combining sample estimates and estimates based on auxiliary data ("auxiliary estimates") would be simply to average them. For example, when 1970 population estimates for the 11 PSU's in the Current Population Survey with populations of more than 2 million were obtained by averaging sample estimates from the October 1969 and January, April, July, and October 1970 surveys, the average error was 2.0 percent. These sample estimates could be averaged with selected auxiliary estimates to produce an optimal combination, perhaps using empirical Bayes techniques. Such averaging would be a significant step, since these 11 areas included just over 25 percent of the nation's 1970 population. A second way of combining sample data with auxiliary estimates is by the regression-sample data procedure originally formulated by Hansen et al. (1953) and applied to the population estimation problem by Ericksen (1974). Here a regression equation using auxiliary estimates and other auxiliary information is computed by using sample estimates, in this case obtained from the CPS, of the variable in question (the dependent variable). If only auxiliary estimates are used, this regression equation estimates the optimal weighting allocation among the auxiliary estimates. Examples of how 1973 and 1975 population estimates could have been improved by such a procedure are given below.

Disregarding combinations of procedures for the moment, it is clear that the first procedure should be favored whenever the most recent decennial census provides more accurate estimates of current population or income than do available postcensal estimation procedures. The accuracy criterion or loss function needs to be specified, but if it is squared relative error, then the procedure should be favored when the variability of rates of change is less than the mean squared relative error of available estimating procedures. [13] For example, if one wanted population and per capita income estimates for 1971, it seems intuitively reasonable to use the 1970 census counts. For a later year it also seems plausible that 1970 census counts might be more reasonable for some units, particularly those with small populations and few data. Moreover, there may be situations in which the first procedure would be favored for estimates for income but not for population.

The second procedure would be used in two circumstances. The first is when sample data are sufficient to provide accurate estimates for local areas (as in the case described immediately above). The second is when sets of areas with common characteristics could be combined and a common estimate formed. For example, one could use CPS data to obtain an estimate of per capita income for "central cities under 250,000 population in the Northeast." Different categories of local areas could be considered, and the accuracy of estimates could be assessed by variance computations and other techniques.

The third procedure, which is the one used by the Census Bureau, should be chosen when the auxiliary information is available and there is so much variation among local areas in a category that combined sample estimates have large errors. The auxiliary information is available from vital statistics, school enrollment data, income tax records, and various other sources, which vary by state. The problem with this set of procedures is that there is no satisfactory way of evaluating their accuracy except by conducting special censuses. The Census Bureau and the state agencies that conduct such tests usually use evaluations from preceding intercensal periods to verify accuracy. This can be misleading when relationships among variables change from one time period to another.

5.2b UNIFORMITY

The specific comments in this section pertain to population estimation, but the underlying ideas also apply to the generation of income estimates. Although there are some variations by state, the general method used to produce preliminary estimates of county populations is to compute an equally weighted average of the AR and CM II estimates. [14] This

---

[13] In this instance, variability of rates of change should be measured by the average squared change in population, with the change expressed as a proportion of the true current population.

[14] As was noted in Chapter 1, these are the estimates used for determining general revenue sharing allocations; see Appendix A, section 3.1.

average is then subtracted from the corresponding average for the preceding year and added to the revised estimate for the preceding year, which is an equally weighted average of the AR, CM II, and RC estimates. Because it takes longer to obtain the requisite data for the ratio-correlation estimates, these are not available for the preliminary estimates.

The same procedure is used for all counties, whether they are big, small, or rapidly or slowly growing and regardless of the age structure or other demographic characteristic of the population. For subcounty areas, AR estimates are used, since the data for computing the other estimates are not consistently available. Thus while a given procedure may be particularly good for larger areas but another may be particularly good for smaller areas, one or the other of the procedures, or an average of them, is applied to all areas because the Census Bureau uses a uniform procedure throughout.

Several features of the Census Bureau's approach bear examination. One of these is the use of equal weights for averaging the auxiliary estimates. If auxiliary estimates are of unequal accuracy, then unequal weights can produce more accurate estimates than equal weights. As an extreme example, in the most exhaustive test of county estimates conducted by the Bureau to date (in which 2,586 county estimates in 42 states were compared with the 1970 census), it was found that the RC procedure, most accurate among the four procedures tested, gave better results (in terms of average percent difference) than any equally weighted average of two, three, or four estimates. (See Bureau of the Census (1973b, Table C), but note that the term "regression" there refers to the ratio-correlation method.)

A second feature that bears examination is the uniformity constraint. By relaxing this constraint the Bureau could improve the accuracy of its estimates. There are four ways of relaxing uniformity:

1. Different kinds of data could be used for making estimates for different local areas in the same state. For example, different regression equations (using different sets of independent variables) could be applied for different counties within a state.

2. Counties from different states but with comparable data sources could be estimated by a single regression equation (as was done by Ericksen (1974)). Even among counties with comparable data series, separate regressions might be determined for counties differing according to region, size, rate of growth, age structure, or other characteristic.

3. Different methods may be used for different local areas. For example, additional data sources are available for many large cities, so that alternative estimates could be prepared. These alternative estimates

could be averaged with the administrative records method estimates currently used by the Bureau for subcounty areas.

4. Different mixes (weighted averages) of methods could be used for different local areas. For example, the accuracy of component method II estimates drops more rapidly than does that of the administrative records method as population size decreases. When component method II and administrative records method estimates are averaged for counties, the weight assigned to the component method II estimates could be smaller for small areas that for large ones.

5.2c EVALUATION OF THE CENSUS BUREAU'S APPROACH TO WEIGHTING ESTIMATES

This section evaluates the use of equal weights to compute 1975 population estimates for counties. The relative accuracy of alternative weighting schemes, obtained by using the regression-sample data procedure (Ericksen, 1974; Fay, 1979; Gonzalez and Hoza, 1978) can be judged against 130 special county censuses.[15]

In this analysis we evaluate the use of equal weights by comparing the accuracy of the estimates of 1975 population provided by equally weighted averages of 1975 AR and CM II estimates with the accuracy of differentially weighted averages of these estimates. In 1975 the Bureau's preliminary county estimates were derived from the average of administrative record and component method II estimates.[16] If we define X_1 as the AR estimate and X_2 as the CM II estimate, the Bureau's method can be written as $\hat{Y} = .50X_1 + .50X_2$. Four other auxiliary estimates were available for counties in all states: the respective 1975/1970 ratios of the numbers of Medicare recipients, numbers of school children, numbers of income tax exemptions, and income tax returns. Our evaluation consisted of selecting the best combination of these six auxiliary estimates, using regression with the PSU estimates of population growth (from the CPS) serving as the dependent variable to compute weights.

All simple squared correlations (r^2) and multiple squared correlations

---

[15] The computations for this test were carried out at our request by David Word at the Population Division of the Census Bureau.

[16] As was described in section 1.2a, the preliminary county estimates for 1975 were derived as the sum of the revised estimate for 1974 plus the equally weighted average of two estimates of change from 1974 to 1975, obtained as the difference between the 1975 and 1974 AR estimates and the difference between the 1975 and 1974 CM II estimates. (For preliminary county estimates in some states, the difference between the 1975 and 1974 estimates by a third, locally used, method was averaged equally with the differences in the AR and CM II estimates.)

(R^2) for each of the 15 pairs of auxiliary estimates are presented in Table 5.1. No combination of three estimators produced a higher multiple correlation than the best pair of estimators, administrative records and component method II. If one looks first at the multiple correlations, AR and CM II explained 27.4 percent of the variance in the CPS estimates. This is scarcely better than the 27.2 percent of variance explained by the AR estimate alone, however. The small size of this improvement is explained by the similarity of the simple correlations with the CPS estimates ($r = .522$ for administrative records and .504 for component method II) and the extremely high correlation ($r = .940$) between these two auxiliary estimates. We also note that the observed correlations between the auxiliary estimates and the CPS estimates are shrunk toward zero because much of the variance of the CPS estimates arises from random sampling error and hence cannot be explained by the auxiliary estimates.

Because the multiple correlation is so close to the simple correlation, the choice of "best" estimate is ambiguous, but preference could be given to the two-variable equation because of the observed increase in explained

TABLE 5.1   Correlations of Auxiliary Population Estimates With Sample Population Estimates for PSU's From the CPS, 1975

	Simple Correlations (Squared)	Variables	Multiple Correlations (Squared)
X_1, administrative records	.272	X_1, X_2	.274
X_2, component method II	.254	X_1, X_3	.272
X_3, ratio of IRS exemptions	.252	X_1, X_4	.273
X_4, ratio of IRS returns	.246	X_1, X_5	.273
X_5, ratio of school enrollment	.181	X_1, X_6	.272
X_6, ratio of Medicare recipients	.166	X_2, X_3	.263
		X_2, X_4	.267
		X_2, X_5	.255
		X_2, X_6	.258
		X_3, X_4	.253
		X_3, X_5	.252
		X_3, X_6	.253
		X_4, X_5	.257
		X_4, X_6	.246
		X_5, X_6	.222

Note: The PSU's from the CPS were weighted according to the size of the stratum being represented: hence the larger self-representing PSU's had the largest weights.

SOURCE: Computations from the Bureau of the Census provided by David Word (private communication, March 23, 1979).

variance and because the presence of within-PSU sampling error did damp the observed correlations. Computing two regression equations, one using the AR estimate as the single independent variable and the other using AR and CM II estimates in a two-variable equation, we obtain

$$\hat{Y} = .040 + .976X_1 \qquad\qquad r^2 = .272$$

$$\hat{Y} = .027 + .766X_1 + .223X_2 \qquad R^2 = .274.$$

In both cases, county estimates were computed for all counties in the United States. Because the sum of the county estimates was slightly greater than the estimated national population, each estimate was multiplied by 0.985 to make the sum of county estimates agree with the national total.

Table 5.2 shows a comparison of the 130 estimates for counties that had special censuses with their enumerated populations. Notice that the actual results support our predictions based on the correlation and regression analysis. The two-variable regression equation produced the best results, though not by much, with the one-variable regression equation coming in second, giving nearly identical results to the administrative records estimate alone. The equally weighted average of AR and CM II estimates was less accurate overall than the AR estimate alone or either of the regression estimates. Here, use of equal weights detracts from overall accuracy, a result similar to that observed for 1970 when the ratio-correlation estimates were compared to all possible equally weighted averages of four estimates (Bureau of the Census, 1973b, Table C).

It will be noted that the CM II estimates decreased the accuracy most for small counties. This reinforces our suggestion that the Bureau would do well to weight procedures differently for different types of counties.[17] Component method II does well for large counties and can give improved results to those obtained for administrative records alone. The computation of separate CPS-based regression equations for large and small counties might provide guidance on how to produce such stratified estimates, but to our knowledge such experimentation has not been done.

[17]The reader may note that the term "weighted average" is being used loosely, to include both use of negative weights for variables and use of the constant term in the average. The weighted averages may be constructed to satisfy various constraints—e.g., no constant term, nonnegative weights, or weights summing to 1.0—but we do not find compelling motivation for these constraints. Fay (1979) notes an analogy between the sum of the weights being less than 1.0 and the shrinkage phenomenon arising in Stein-James estimators. We also note that although least-squares is the criterion used here to estimate the weights, other criteria pertaining to alternatives (discussed in section 3.1) may also be used.

TABLE 5.2   Average Percent Differences for Alternative 1975 Population Estimates Tested Against 130 County Censuses

	All Counties	Counties With 1975 Populations of		
Type of Estimator		100,000+	5,000 to 100,000	Under 5,000
Component method II alone	6.42	2.31	5.41	10.80
Administrative records alone	4.14	2.03	3.87	6.08
Equally weighted average of component method II and administrative records	4.32	1.79	4.03	6.62
Regression, one variable (AR)	4.12	2.03	3.78	6.14
Regression, two variables (AR, CM II)	4.01	1.82	3.73	6.03

Note: Estimates refer to July 1, 1975. The county censuses against which the estimates were compared were taken between July 1, 1974, and December 31, 1976, and linearly interpolated or extrapolated to July 1, 1975, from the April 1, 1970, counts. Percent difference for each county equals estimate (as of July 1) *minus* adjusted special census count (interpolated or extrapolated to July 1, 1975), expressed as a percent of the adjusted special census count. Average percent difference was calculated as the arithmetic mean of percent differences *disregarding sign.*

SOURCE: Computations from the Bureau of Census provided by David Word (private communication, March 23, 1979).

As was noted above, the fallacy of arbitrary averaging was also demonstrated by 1970 data. For that time, four auxiliary estimates—ratio-correlation, component method II, composite, and vital rates—were all available. The average percent difference of the ratio-correlation estimates was 4.6 percent, less than the average percent difference for any equally weighted average of two, three, or four estimates (Bureau of the Census, 1973b, Table C). When those four auxiliary estimates were added to three ratios—births, deaths, and school enrollment in the same kind of exercise that we have just reported—the best combination of variables was composed of births, deaths, school enrollment, and the ratio-correlation estimates. For this, $R^2 = .428$, and the regression equation was

$$\hat{Y} = .085 + .745 \text{ (ratio-correlation)} + .214 \text{ (school enrollment)} + .045 \text{ (deaths)} - .097 \text{ (births)}.$$

When the estimates obtained from this equation were made for 2,586 counties in 42 states, the average percent difference obtained was 4.2

percent, and the number of large errors (10 percent or more) was 194, compared with 264 large errors obtained from ratio-correlation alone. No improvements were obtained by adding the other three auxiliary estimates to the equation or by substituting them for births, deaths, or school enrollment.

5.2d IMPROVING THE PRECISION OF SAMPLE DATA

Data from surveys such as the CPS and the Annual Housing Survey should be used more in the postcensal estimation program, both in producing and in evaluating estimates. The usefulness of CPS data for postcensal population estimates could be further enhanced by certain changes in the CPS design. One recent design change made by the Census Bureau is the monthly collection of age, race, and sex data for each household member. This change will allow more data to be pooled to provide better yearly sample estimates. The Panel suggests four additional changes:

1. Data on central city or suburban location as well as identification of county of residence should be collected. These data would facilitate the computation of separate regression equations for different types of counties as well as separate equations for central cities for other types of local areas.

2. The CPS is currently designed to minimize the variances of national and state unemployment and employment estimates. Research should be done to see if the CPS could be redesigned to improve the accuracy of population and income estimates for local PSU's without substantially increasing the variances of the state employment and unemployment estimates.

3. Within-PSU samples should be selected in such a way as to facilitate estimation of within-PSU variance, provided such redesign does not substantially increase the variances.

4. If the CPS sample were enlarged, particularly in non-self-representing areas, precision of the estimates would be increased; however, this would involve substantial expense and may not be practicable.

REFERENCES

Advisory Commission on Intergovernmental Relations (1971) *Measuring the Fiscal Capacity and Effort of State and Local Areas.* Washington, D.C.: Advisory Commission on Intergovernmental Relations.

Bureau of the Census (1973a) *Characteristics of the Population, Vol. 1, Census of Population: 1970.* Washington, D.C.: U.S. Department of Commerce.

Bureau of the Census (1973b) *Federal-State Cooperative Program for Local Population Estimates: Test Results—April 1, 1979.* Current Population Reports, Series P-26, No. 21. Washington, D.C.: U.S. Department of Commerce.

Bureau of the Census (1974) *Estimates of the Population of States With Components of Change: 1970-1973.* Current Population Reports, Series P-25, No. 520. Washington, D.C.: U.S. Department of Commerce.

Bureau of the Census (1976) *Estimates of the Population of States With Components of Change: 1970-1975.* Current Population Reports, Series P-25, No. 640. Washington, D.C.: U.S. Department of Commerce.

Bureau of the Census (1978) *Population Estimates by Race, for States: July 1, 1973 and 1975.* Current Population Reports, Series P-23, No. 67. Washington, D.C.: U.S. Department of Commerce.

Bureau of the Census (1979) *Estimates of the Population of the United States and Components of Change: 1940-1978.* Current Population Reports, Series P-25, No. 802. Washington, D.C.: U.S. Department of Commerce.

Bureau of the Census (1980) *Population and Per Capita Money Income Estimates for Local Areas: Detailed Methodology and Evaluation.* Current Population Reports, Series P-25, No. 699. Washington, D.C.: U.S. Department of Commerce.

Bureau of the Economic Analysis (1977) *Local Area Personal Income, 1970-1975,* Vol. 1, Washington, D.C.: U.S. Department of Commerce.

Ericksen, E. P. (1974) A regression method for estimating population change for local areas. *Journal of the American Statistical Association* 69(348):867-875.

Ericksen, E. P. (1975) Population Estimation in the 1970s: The Stakes are Higher. Institute for Survey Research, Temple University, Philadelphia, Pa.

Fay, R. E. (1979) Some recent Census Bureau applications of regression techniques to estimation. Pp. 185-194 in National Institute on Drug Abuse, Research Monograph 24, *Synthetic Estimates for Small Areas: Statistical Workshop Papers and Discussion.* Washington, D.C.: U.S. Government Printing Office.

Fay, R. E., and Herriot, R. (1979) Estimates of income for small places: An application of James-Stein procedures to census data. *Journal of the American Statistical Association* 74(June):269-277.

Ferreira, J. (1978) Identifying equitable insurance premiums for risk classes: An alternative to the classical approach. Pp. 74-120 in *Automobile Insurance Risk Classification: Equity and Accuracy.* Boston, Mass.: Division of Insurance.

Gonzalez, M. E., and Hoza, C. (1978) Small area estimation with application to unemployment and housing estimates. *Journal of the American Statistical Association* 73(361): 7-15.

Hansen, M. H., Hurwitz, W. N., and Madow, W. G. (1953) *Sample Survey Methods and Theory,* Vol. 1. New York: John Wiley and Sons.

Lowe, T., Walker, J., and Weisser, L. (1974) *Evaluation of Estimates Based on Income Tax Returns.* Staff Document No. 15. Population Studies Division, Office of Program Planning and Fiscal Management, State of Washington.

Mann, E. (1978) Problems with data in selected formula funded programs as applied to New York City. Pp. 114-119 in *1978 Proceedings of the Section on Survey Research Methods of the American Statistical Association.* Washington, D.C.: American Statistical Association.

Mosteller, F., and Tukey, J. W. (1977) *Data Analysis and Regression: A Second Course in Statistics.* Reading, Mass.: Addison-Wesley.

Namboodiri, N. K., and Lalu, N. M. (1971) The average of several simple regression estimates as an alternative to the multiple regression estimate in postcensal and intercensal population estimation: A case study. *Rural Sociology* 36(2):187-194.

Office of Federal Statistical Policy and Standards (1978) *Statistical Policy Working Paper 1: Report on Statistics for Allocation of Funds.* Prepared by the Subcommittee for Allocation of Funds, Federal Committee on Statistical Methodology. Washington, D.C.: U.S. Department of Commerce.

Ono, M. (1972) Preliminary evaluation of 1969 money income data collected in the 1970 census of population and housing. Pp. 390-396 in *1972 Proceedings of the Social Statistics Section of the American Statistical Association.* Washington, D.C.: American Statistical Association.

Sahai, H. (1979) A bibliography on variance components. *International Statistics Review* 47(2):177-222.

Schultz, T. W. (1964) *Transforming Traditional Agriculture.* New Haven, Conn.: Yale University Press.

Stanford Research Institute (1974) *General Revenue Sharing Data Study,* Vols. III and IV. Menlo Park, Calif.: Stanford Research Institute.

Voss, P. R. (1978) Precision in the Wisconsin small-area population estimation system: A study in variation in estimates as length of time from the estimation year increases. Technical Series 70-3, Applied Population Laboratory, College of Agriculture and Life Sciences, University of Wisconsin, Madison.

PART

III APPENDICES

Postcensal Population Estimation Methods of the Census Bureau

BRUCE D. SPENCER and
CHE-FU LEE

NOTE: The descriptions in this paper are based on the authors' understanding of the methods used by the Bureau of the Census to estimate population over the period 1970-1977.

The Census Bureau is continually refining its procedures in (usually) minor ways, and such changes are noted wherever possible. Nevertheless, because of the ongoing modifications and because of the great complexity of the methodology the methods practiced by the Census Bureau and as described below may differ in minute details.

The generous and indispensable assistance of Census Bureau staff is gratefully acknowledged, in particular, that of David Word, Frederick Cavanaugh, Mary Kay Healy, Jennifer Peck, Richard Irwin, Jerome Glynn, David Galdi, Joseph Knott, Edward Hanlon, Marianne Roberts, Barbara van der Vate, Richard Engels, Louisa Miller, Sharon Baucom, Frances Barnett, and Joel Miller. They discussed the Bureau's methodology with the authors, and several of them reviewed earlier drafts of this appendix. Final responsibility for the accuracy of the descriptions rests with the authors.

CONTENTS

INTRODUCTION AND OVERVIEW

Postcensal estimates refer to a date (past or current) following a decennial census and use that census and possibly earlier censuses as a point of departure. To understand postcensal population estimation methods for small areas, it is necessary first to understand those for the larger units of population—counties, states, and the nation as a whole. The reason for this is that the Census Bureau prepares its postcensal estimates by first making the national estimate. Then estimates for the 50 states and the District of Columbia are made and controlled (forced to sum) to the national total. Subsequently, county estimates are controlled to a state total, and subcounty estimates to a county total.

Essentially, two kinds of methods (component and regression) are used. The component method calculates separately three elements of population dynamics: net natural increase (number of births minus deaths), migration (net inmigration, including immigration), and changes in "special populations" not reflected in symptomatic data, namely, group quarters populations. These individual components are then aggregated to yield an estimate of population change.

In the regression method, equations are constructed to relate observed population changes to observed changes in other "symptomatic" data that are available and considered relevant to population changes. Subsequent observed (postcensal) changes in symptomatic data are then transformed by the equations to yield estimates of postcensal changes in population.

Postcensal estimates of the total U.S. population are made using a component method. This procedure is described in Part 1 of this appendix.

The state population estimates are derived by averaging the results of three methods: component method II (CM II), administrative records method (AR), and ratio-correlation method (RC). Component method II and administrative records are both variations of component methods. They differ only in estimation of net migration: CM II relies on changes in school enrollments, while AR uses matched individual federal income tax returns and treats net immigration separately. Ratio-correlation is a regression method. State procedures are discussed in Part 2.

County estimates (discussed in Part 3) are generally produced from methods similar to those used in state estimates. However, in some states, counties use additional information, such as data on drivers license registrations or new housing units.

Finally, methods for estimating subcounty populations are described in Part 4. With a few exceptions the procedures are similar to the administrative records method used at the state level.

Estimation methods are described here in the statistical tradition, in that parameters are introduced and the objective of the estimation procedures is accurate estimation of the parameters. A difficulty in describing the estimation procedures in this way is the lack of well-defined stochastic or demographic accounting models underlying the procedures. This occurs because the descriptions written by the Bureau of the Census outline their procedures and their objective but fail to specify in detail the models underlying the procedures.

A statistical model should be consistent: if each parameter in the model were perfectly estimated, the objective described by the model (here, total population) would be perfectly estimated. It is not permissible to omit parameters entirely, even if data to estimate them are not available. For example, a model for postcensal change in total U.S. resident population should not exclude a component for change in the number of "illegal" immigrants, even though satisfactory data may be lacking. This component does not have to appear as a separate entity—it may be incorporated into one or more other components—but it must not be omitted entirely. While it is permissible to use estimators that fail to coincide with the parameters as to geography or time of reference, the model itself must be consistent and well specified.

The deviation of an estimate from its parameter is referred to as error. The sources and structure of error will be discussed below for the various postcensal estimation methods used by the Bureau of the Census. It should be recognized that a major, for many areas *the* major, source of error in the estimate of total postcensal population is undercoverage (undercount) in the decennial census. On the other hand, undercoverage (for small areas) is a minor component of error for the estimates of postcensal *change* in population. For this reason, discussion of the sources and structure of error will generally omit undercoverage as a source of error.

PART 1   U.S. POPULATION

1.1 INTRODUCTION

The resident population of the United States includes residents of the 50 states and the District of Columbia. It does not include residents of Puerto Rico and the outlying areas under U.S. sovereignty or jurisdiction, armed forces stationed in foreign countries, and other American citizens residing outside the United States. Postcensal estimation of this total resident population during the period 1970–1977 is described below with respect to methodology (section 1.2), sources of data and errors (section 1.3), and error structure (section 1.4). Apportionment of the estimated total by age, race, and sex is discussed in section 1.5.

1.2 METHODOLOGY

The Bureau of the Census makes postcensal estimates of the U.S. resident population by estimating components of population change since the previous decennial census (1970). These components, to be discussed in the following sections, include natural increase, net immigration of U.S. armed forces from abroad, and net civilian immigration.

The change in total population since the 1970 census is estimated by the sum of the three components of change. The estimate of postcensal population is then obtained by adding the estimate of change to the 1970 population count.

1.2a *Natural Increase*

Natural increase equals the number of births minus the number of deaths. The National Center for Health Statistics (NCHS) provides reports of these numbers. Until 1970 (1960) adjustments were made for estimated under-registration of births (deaths). These adjustments are no longer made because the amount of underregistration is believed to be small and because of the difficulty of correctly apportioning the imputed births to subnational areas.

1.2b *Net Immigration of Armed Forces From Abroad*

This component is estimated by the following total:

- (a) number of armed forces abroad in 1970
- −(b) number of armed forces abroad on the estimate date
- −(c) number of deaths to armed forces abroad since previous census
- +(d) net change in number of recruits from Puerto Rico who are with the armed forces.

These numbers are obtained from the Directorate of Information of the U.S. Department of Defense and from the Army, Navy, Air Force, Marine Corps, and Coast Guard.

1.2c *Net Civilian Immigration*

Net civilian immigration is estimated by the following total:

- (a) alien immigration
- +(b) parolee immigration
- +(c) net arrivals from Puerto Rico

+(d) net movement of civilian citizens associated with the U.S. government

−(e) other net emigration (including migration of U.S. citizens and aliens not included in (c) and (d) above).

Alien immigrants are those nonrefugee aliens accepted for permanent residence by the Immigration and Naturalization Service (INS). The INS classifies an individual as an immigrant when it grants permanent residence status. Since this does not necessarily coincide with the time of physical entry into the United States, the Census Bureau reallocates immigrants for whom the data are available to date of entry. Most Cubans and Indochinese who change their status from nonimmigrant to permanent resident alien can be reallocated to date of entry, but generally, other nonimmigrant aliens cannot be reallocated when they adjust to immigrant status. Many other individuals (notably students) who enter with nonimmigrant visas later adjust to immigrant status, but these people are not reallocated to date of entry.

The Bureau of the Census does not attempt to include illegal alien migrants or aliens temporarily residing in the United States in the estimate of net immigration. The latter group includes aliens with temporary visas (students, visitors, diplomats) and numerous agricultural workers from Mexico and the British West Indies working under special contract.

The classification "parolee" refers to nonimmigrant aliens other than permanent resident aliens who are allowed to remain in the United States permanently. Parolees consist almost entirely of refugees from Indochina, Cuba, Hong Kong, and communist countries of eastern Europe. The alien immigration figures do not include the parolees. Counts of immigrant parolees are provided by the Immigration and Naturalization Service and the Task Force for Indochina Refugees of the U.S. Department of Health, Education, and Welfare (HEW).

Net arrivals from Puerto Rico are estimated on the basis of passenger statistics. The Puerto Rico Planning Board collects data from air and sea carriers on passengers entering and leaving Puerto Rico. The difference between the number of departures from Puerto Rico and the number of arrivals to Puerto Rico is used to estimate the net migration from Puerto Rico to the United States. The implicit assumption is that the migration between Puerto Rico and countries other than the United States is insignificant. To reduce the fluctuations that can arise from the seasonal movement of travelers, passenger counts are smoothed by a 12-month running average.

Information about civilian citizens migrating to or from the United

States is most reliable for those individuals affiliated with the U.S. government. This group includes overseas civilian citizen employees of the federal government as well as overseas citizen dependents of federal employees and servicemen. Data from the U.S. Department of Defense and *Federal Civilian Work Force Statistics* are used to estimate the total change in the number of civilian citizens affiliated with the U.S. government, and their dependents, who are overseas during the postcensal period. The natural increase (births, estimated from reports of Department of Defense hospitals) over the period is subtracted from this total change. The negative of the residual is taken as the estimate of net civilian citizen immigration over the period. Deaths in the overseas civilian citizen population are ignored. Also ignored are civilian citizens overseas who leave federal employment but remain overseas and civilian citizens living overseas who accept federal employment.

"Other net emigration" refers primarily to persons not affiliated with the federal government who move from the United States to a foreign country. Since 1957 no statistics have been collected on the number of persons who have permanently moved out of the United States. Estimates are based on 1960–1970 data on overseas payments from the Social Security Administration and data reported to the United States by foreign countries on numbers of immigrants into these countries. The Census Bureau assumes that the level of emigrants has remained constant since 1970 (see Warren and Peck, 1975).

1.3 SOURCES OF DATA AND ERROR

The net civilian immigration component is subject to greater error in estimation than natural increase or net immigration of armed forces for the following reasons:

1. Illegal alien migrants are not identified as such, and no one knows what fraction of them are counted as residents. Good estimates of the extent of such undocumented alien immigration are lacking, and the magnitude of this error is difficult to estimate.

2. Net arrivals from Puerto Rico are estimated from airline passenger data. The determination of a small net flow from large gross flows in and out of Puerto Rico of approximately equal magnitude (residual process) is not conducive to accurate measurement. Net immigration from other U.S. possessions is not estimated at all.

3. Estimates of emigration are markedly understated because statistics on permanent arrivals from the United States are provided by few countries, and when they are available, data are generally poor and variable in

coverage. The official estimates of emigration in the 1970s are 36,000 per year, but Warren and Peck (1975), using demographic analysis, estimated that over 100,000 persons from the foreign-born population emigrate each year. The total number of emigrants, native and foreign born, is even higher.

Error in the estimate of population at the time of the census is very significant for estimating total population. Net undercoverage in the decennial census has received substantial discussion (see Bureau of the Census, 1973a). Recognition of the undercoverage problem has led to the development and use of the inflation-deflation method, discussed below in section 1.5. This method reduces the impact of undercoverage on the estimates of postcensal population change of age groups but does not affect the estimates of postcensal change for the population as a whole.

Birth underregistration is believed to be small (see Bureau of the Census, 1973c).

1.4 STRUCTURE OF ERRORS

The errors in the national components of change and in the total national population estimates are important for subnational estimates because the small-area estimates are constrained by the national estimates in various ways. For example, the national estimate constrains the subnational estimates by broad age groups. Postcensal estimation of state populations involves separate estimation of the population aged 65 and over; these estimates in each state are scaled so they sum to the national estimate of population 65 and over. Since age composition varies from state to state, error in the estimate of the national population 65 and over affects states differentially.

1.5 APPORTIONMENT OF NATIONAL POPULATION BY AGE, RACE, SEX

Postcensal estimates of national population by age, race, and sex are obtained by using a method called inflation-deflation. First, the 1970 estimate of total population including military overseas is adjusted for estimated census undercoverage by age-race-sex class. The undercoverage rates are based on set D of the Bureau of the Census (1973a). The armed forces are assumed to be completely counted. Second, births (but not deaths) are adjusted for underregistration by race-sex. This is the "inflation" part of the procedure.

Next, the components of population change are broken down into age-race-sex categories. The following methods and data sources are used:

1. Resident births, deaths: National Center for Health Statistics (NCHS) gives data on age-race-sex.

2. Deaths to armed forces abroad: Data on total military deaths are obtained from each branch of the armed forces. The Census Bureau estimates deaths to armed forces overseas by assuming that the proportion of these latter deaths for each state is the same as the proportion of total military deaths for the state.

3. Alien immigrants: Immigration and Naturalization Service has data on age, sex, and country of birth. Race is apportioned according to the pattern observed in the 1970 census for immigrants by country of birth over the period 1965–1970.

4. Parolees and refugees: Cubans were assumed all white, with age-sex distributions the same as Cubans in point 3. Classification of Vietnamese was based on counts of the HEW Task Force on Indochina Refugees.

5. Net arrivals from Puerto Rico: Puerto Ricans were assumed all white and are currently classified by age-sex according to the age-sex distributions for Puerto Ricans living in the United States in 1970 (based on census data). In the years prior to 1977 the distributions were based on surveys of inmigrants and outmigrants from Puerto Rico.

6. Civilian citizen immigrants affiliated with the United States: These persons are distributed according to the observed 1970 census age-race-sex distribution of this population overseas.

7. Other emigrants: Social security beneficiaries are assumed to be over 65 and are classified by race-sex on the basis of social security data. Canada provides the Bureau of the Census with age-sex-race distributions of American migrants to Canada. Migrants to other countries are assumed to have the same age-race-sex distributions as migrants to Canada.

The final step consists of "deflating" the estimates of each age-race-sex group by multiplying each estimate by the corresponding undercoverage rate in the 1970 census. The same rates (set D of the Bureau of the Census, 1973a) are used to deflate as were used to inflate, but the rates are applied to *age groups* rather than cohorts. For example, if R_5 and R_{10} were the estimated 1970 undercoverage rates for white male children aged 5 and 10 in 1970, then in estimating the 1975 population the 1970 base population of white males aged 5 would be inflated by $(1 - R_5)^{-1}$. However, the estimate of persons aged 10 in 1975 based on the cohort aged 5 in 1970 would be deflated by $1 - R_{10}$.

After each age-race-sex class is deflated, further adjustment forces the total over subgroups to equal the national total obtained without inflation or deflation. The adjustment is necessary because inflation-deflation is consistent for age groups but not for cohorts. Finally, the overseas military component is subtracted from the national total.

The rationale of the inflation-deflation method flows from the assumption that undercoverage rates for age groups are stable over time. The ultimate purpose of the method is to provide accurate estimates of postcensal population change by age, and the strategy for achieving this goal is to preserve in the postcensal estimates the 1970 undercoverage structure for age groups. As a result, the postcensal age distribution does not reflect the 1970 undercoverage structure for cohorts. For example, the estimated numbers for immigrants not present in 1970 (hence not subject to undercount) are nonetheless deflated. In the absence of inflation or deflation, direct application of the age-specific rates of change (birth, death, migration) to the various age groups would preserve in the postcensal estimates the 1970 undercoverage structure for cohorts but would not preserve the undercoverage structure across age groups.

PART 2   STATE POPULATIONS

2.1 INTRODUCTION

The Census Bureau derives postcensal estimates of state populations by averaging the results of three methods: component method II (CM II), ratio-correlation method (RC), and administrative records method (AR).[1] These methods have the following features in common: (1) Current data are used to estimate population change since the previous census (or since a recent postcensal estimate). (2) Change in the 65 and over population is estimated separately, by using Medicare enrollment data. (3) Change in the population living in group quarters is treated separately.

Both CM II and AR are component methods. In using these methods,

postcensal population = base population
+ births
− deaths
+ net migration
+ changes in group quarters population.

---

[1] Except for the provisional estimates, which are typically based on just two methods. For example, the Census Bureau made the provisional estimates by adding to the revised 1975 estimate the average change between 1975 and 1976 for component method II and a two-variable ratio-correlation estimate. In addition, component method II was not used for estimating Alaska population beginning with 1975 (see Bureau of the Census (1978b) for discussion).

The principal components of population change are net natural increase (births minus deaths) and net migration.[2] Special populations, i.e., those living in group quarters, are handled separately because changes in size of the special populations are not adequately reflected in the data used for the rest of the population. State special populations have included residents of military barracks, large Job Corps centers, institutions (mental hospitals, correctional facilities, etc.), college dormitories, and (for 1975) Vietnamese in resettlement camps.

Both AR and CM II estimate net natural increase similarly and migration differently. To estimate migration, CM II uses school enrollment data for internal migration and immigration, while AR matches Internal Revenue Service individual income tax returns for internal migration and treats immigration separately.

In the ratio-correlation method (RC), regression equations are used to relate population changes to changes in symptomatic data or indicator variables (see Morrison, 1971; Purcell and Kish, 1979). The RC method proceeds in two steps: (1) construction of a regression equation using data from a base observation period and (2) use of this equation to estimate postcensal population from current symptomatic data.

To describe the methods in detail, it will be useful to develop notation, which will be introduced as needed, with a summary appearing as a special note at the end of Appendix A. A convention in use here is that a person who would be 65 or older on the estimate date is "elderly"; all other persons are "young." Methods CM II, RC, and AR will each be discussed in turn, with attention to methodology, sources of data and error, and error structure.

2.2 CM II METHODOLOGY

2.2a *Introduction and Overview*

It is convenient to let T refer to time in years, with $T = 0$ the time of reference of the previous census and $T = t$ the time of reference of the present estimate. The interval $(T_1, T_2]$ is the period since time T_1 up to and including time T_2.

---

[2] From 1970 to 1975 the change in the total population of all the states increased about 8.6 percent from births, decreased about 5.0 percent from deaths, and increased 1.2 percent from net migration. These rates vary substantially among the states with respect to net migration, ranging from −8.1 percent in the District of Columbia to +20.8 percent in Florida (see Bureau of the Census, 1976).

The model underlying CM II is[3]

RESPOP(t) (postcensal population)
= POPY(0) (April 1, 1970, young population)
+ BIR(0, t) (births)
− DEAY(0, t) (deaths to young)
+ NGQMIGY(0, t) (net migration of non-group quarters young persons)
+ GQPOPY(0, t) (net change of group quarters young)
+ NETMOVY(0, t) (net movement of young from military group quarters to non-group quarters)
+ POPE(t) (elderly population),

where (the following notation refers to a particular state)

RESPOP(T)   resident population at time T;
POPE(T)   resident elderly population at time T;
POPY(T)   resident young population at time T;
BIR(T_1, T_2)   number of resident births in $(T_1, T_2]$;
DEAY(T_1, T_2)   number of resident deaths to young in $(T_1, T_2]$;
NGQMIGY(T_1, T_2)   number of young persons newly taking up non-group quarters residence in the state over interval $(T_1, T_2]$ minus the number of young moving out from non-group quarters in the state either to another state or to group quarters in the state over interval $(T_1, T_2]$;
GQPOPY(T_1, T_2)   number of young persons newly taking up group quarters residence in the state over interval $(T_1, T_2]$ minus the number of young moving out from group quarters in the state either to another state or into non-group quarters in the state over interval $(T_1, T_2]$;[4]
NETMOVY(T_1, T_2)   excess of young persons moving out of military barracks in the state over those moving into military barracks in the state over $(T_1, T_2]$.

---

[3] The model used by the Census Bureau is slightly different in description but equivalent in operation to that described here. In particular, the Census Bureau estimates the total group quarters young population at time t and adds this to the estimated April 1, 1970, young non-group quarters population (= POPY(0) minus group quarters young on April 1, 1970), plus the other components (BIR(0, t), etc.).

[4] For state population estimates produced in the first half of the decade this component referred to the net movement to the armed forces from the civilian populations rather than to the military barracks population from non-barracks populations.

More notation will be introduced as needed. When it is necessary to be precise in referring to a particular state i, the argument T or T_1, T_2 will be replaced by T; i or T_1, T_2; i, for example, RESPOP(T; i) or BIR(T_1, T_2; i).

2.2b *The Elderly Population:* POPE(t)

To estimate POPE(t), an estimate of the change in the number of elderly is added to the count of POPE(0). The change in the elderly population from time 0 to t is based on the change in the number of Medicare enrollments. Since almost the entire population 65 and over is enrolled in Medicare, the change in number of enrollments in a state reflects both increases from individuals just turned 65 and inmigration of elderly persons and decreases from deaths and outmigration of elderly persons.

The Medicare data base is discussed further in section 3.1. Because time 0 refers to April 1 while the Medicare data refers to July 1, the Medicare enrollments for time 0 are estimated by linear interpolation. Thus, for example, change in Medicare enrollments for a state over the period April 1, 1970, to July 1, 1974, is estimated by

$$\text{MEDCARE}(74) = \{.25\text{MEDCARE}(69) + .75\text{MEDCARE}(70)\},$$

where MEDCARE(x) is the count of Medicare enrollments for the state in year x.

2.2c *Births, Deaths to Young:* BIR(0, t), DEAY(0, t)

Estimates of these two components of natural increase are based primarily on data obtained from state vital statistics offices. These reports of deaths give breakdowns by race but not by age. To estimate DEAY(0, t), it is necessary to differentiate deaths to persons under 65, and for this, National Center for Health Statistics (NCHS) data are used, since NCHS provides age by race breakdowns of total national deaths.

Estimation of DEAY(0, t) will be described for times t, prior to 1979. Beginning with the 1979 population estimates the Census Bureau will estimate deaths to the young directly on the basis of reported deaths by age by state, and the following procedure will no longer be used. Some temporary notation will be useful: let subscripts r, a, i refer to race, age, state and let the argument x refer to the year ending on December 31. Race r takes on two values (white, black and other), as does age a (young, elderly). Consider the notation

$D_{ri}(x)$ reported number of deaths to race r in state i for year x (obtained from state vital statistics offices);

d_{ra} NCHS estimate of the nationwide death rate for persons of race r and age group a over the interval $(0, t]$;

P_{rai} count of race r, age a population of state i on April 1, 1970.

The reported number of deaths to race r in state i over the period $(0, t]$ is denoted by D_{ri}. Because x refers to December 31 while time t refers to July 1 and time 0 refers to April 1, some interpolation is used to obtain D_{ri}. The estimates D_{ri} are also adjusted to the national total. For example, with t referring to July 1, 1973, D_{ri} satisfies

$$D_{ri} = K[.75D_{ri}(70) + D_{ri}(71) + D_{ri}(72) + .50D_{ri}(73)],$$

where K is chosen so that the sum of D_{ri} over states equals the national total.

The estimate of deaths in state i over $(0, t]$ to persons of race r and age a will be denoted by D_{rai}. To obtain this estimate, the estimates D_{ri} are apportioned among the two age groups by using the national death rates:

$$D_{rai} = D_{ri} \frac{d_{ra}P_{rai}}{d_{ry}P_{ryi} + d_{re}P_{rei}},$$

where a takes on the values e (elderly) and y (young).

The estimate DEAY$(0, t; i)$ is then obtained by summing the estimates of the young deaths:

$$\text{DEAY}(0, t; i) = \sum_r D_{ryi}.$$

The estimation of births BIR$(0, t)$ is easier because all newborn are young. Estimates of births provided by a state are nonetheless controlled to national totals. Let

$B_i(x)$ state-provided estimates of births for state i in year x;

B_i unadjusted estimate of births for state i over the interval $(0, t]$;

B NCHS estimate of the total number of births to U.S. residents over the interval $(0, t]$.

The estimate B_i is obtained by interpolation. For example, with t referring to July 1, 1973, B_i satisfied (for the revised estimates)[5]

---

[5] For the provisional estimates, B_i satisfied $B_i = .75B_i(70) + B_i(71) + 1.5B_i(72)$.

$$B_i = .75B_i(70) + B_i(71) + B_i(72) + .50B_i(73).$$

The estimates of BIR(0, t; i) are then obtained by adjusting the B_i to the national total; thus

$$B_i \cdot B/(\Sigma B_i).$$

2.2d *Changes in Special Populations:* GQPOPY(0, t) *and* NETMOVY(0, t)

To estimate the change in special, or group quarters, populations of the young, the Bureau of the Census assumes that there is no net interstate movement of young persons living in group quarters, except for areas controlled by the federal government, including barracks populations of military installations, Job Corps centers in six states, and refugee camps for Vietnamese (in four states in 1975). In estimating substate populations, additional special populations are considered.

The net movement component is estimated on the basis of changes in the size of the total population living in barracks (including those in foreign countries). The total number of persons leaving the barracks is allocated among the states according to the state distributions of preservice residence reported in U.S. Department of Defense records.

2.2e *Non-Group Quarters Migration of the Young:* NGQMIGY(0, t)

An essential element of CM II is the use of school enrollment data to estimate NGQMIGY(0, t). The method will first be sketched and then described in detail.

An estimate of the school-age population for time t is obtained by relating the school-age population to elementary school enrollment at time $T = 0$ (April 1, 1970) and applying this relationship to the school enrollment at time t. This estimate of the school-age population is then compared with the "survivors" of the school-age cohort ("expected" cohort size if there were mortality but no migration in (0, t]). The migration of the school-age population is estimated by the resulting difference between the estimated school-age population and the survivors of the school-age cohort. Dividing the estimated school-age migration by the school-age cohort minus one-half the deaths to the school-age cohort produces an estimate of the school-age migration rate. This estimated school-age migration rate is then adjusted to a migration rate for the young female population. The migration rate for the non-group quarters young population is assumed equal to the rate for young females and then is applied to a base population to yield the estimate of NGQMIGY(0, t).

The following approximations underlie the method: (1) Children start first grade in the calendar year of their sixth birthday. (2) No children fail or skip one of grades 1-8. (3) No children move or drop out during the school year while in grades 1-8. (4) Within each state, the proportion of children aged 6.25-14.24 on April 1, 1970, who are enrolled in grades 1-8, is constant over time. (5) The difference between the average annual migration rates of young females and of children aged 6.25-14.24 remains constant over time in each state. (6) The migration rate for young females in a state equals that of the non-group quarters young population.

The method proceeds in the following manner:

Step I. Obtain school enrollment data directly from a state source to estimate ENROL(0), and ENROL(t), where ENROL(T) is the sum total of all children in the state enrolled in the fall[6] for grades 1-8 in public, private, or special education schools for the school year beginning in the fall of the calendar year preceding time T.

Step II. Estimate school-age population SCLPOP(t) according to

$$\text{SCLPOP}(t) = \frac{\text{SCLPOP}(0)}{\text{ENROL}(0)} \times \text{ENROL}(t),$$

where SCLPOP(T) is the school-age population, precisely, the population aged 6.25-14.24 on April 1 of the calendar year containing time T.

Step III. Scale the estimates of SCLPOP(t) so that they sum to the national estimate (described in Part 1).

Step IV. Estimate "expected" school-age population EXSCLPOP(t), where EXSCLPOP(T) is the "expected" school-age population at time T if there were births and deaths but no migration over the period $(0, T]$. The estimate is made by adjusting the cohort counted in the 1970 census for births and deaths over $(0, T]$. Births and deaths are estimated from reported calendar year vital statistics. To allocate deaths to the school-age population, the national period death rate is applied to the school-age cohort. Denote the deaths to the school-age population over the period $(0, T]$ by SCLDEA(T), denote the number of children born since the last census who attain school age by time T by SCLBIR(T), and denote the size of this cohort at time T by SCLCHT(T). Thus EXSCLPOP(T) = SCHLCHT(0) + SCLBIR(T) − SCLDEA(T).

Step V. Estimate the school-age migration rate SCLMIGRAT(t) according to

$$\text{SCLMIGRAT}(t) = \frac{\text{SCLPOP}(t) - \text{EXSCLPOP}(t)}{\text{SCHLCHT}(0) - \frac{1}{2}[\text{SCLDEA}(t) - \text{SCLBIR}(t)]},$$

[6] For some states, April enrollments are used.

where SCLMIGRAT(T), $T > 0$, is the period net migration rate for population aged 6.50–14.49 at time $T > 0$, over the interval $(0, T]$.

Steps VI–VIII, below, are designed to provide an estimate of the non-group quarters young migration rate NGQMIGYRAT(t), where NGQMI-GYRAT(T), $T > 0$, is the period net migration rate for the non-group quarters young population at time $T > 0$, over the interval $(0, T]$.

To relate the migration rates for the school-age population to the young non-group quarters population, data for the period preceding $T = 0$ are used. Migration rates for young females are computed as an intermediate step to avoid difficulties attendant to military migration during this period.

Step VI. Obtain estimates (using 1970 census data) of the school-age and young female migration rates SCLMIGRAT(0), FEMIGYRAT(0), where SCLMIGRAT(0) is the period net migration rate for population aged 5.00–14.99 at $T = 0$ over the interval $(-5, 0]$ (i.e., over the preceding 5 years) and FEMIGYRAT(0) is the period net migration rate for young females at time $T = 0$ over the interval $(-5, 0]$ (i.e., over the preceding 5 years). The denominators of these period rates are the respective 1970 census populations.

Step VII. Estimate the young female period net migration rate at time t, FEMIGYRAT(t), according to FEMIGYRAT(t) = SCLMIGRAT(t) + [FEMIGYRAT(0) − SCLMIGRAT(0)]($t/5$), where FEMIGYRAT(T), $T > 0$, is the period net migration rate for the young females at time $T > 0$ over the interval $(0, T]$.

Step VIII. Estimate the non-group quarters young period net migration rate NGQMIGYRAT(t) by assuming NGQMIGYRAT(t) = FEMIGYRAT(t).

Step IX. Estimate net non-group quarters young inmigration NGQMIGY(0, t) by multiplying the estimate of the migration rate NGQMIGYRAT(t) by the estimate of the base population. Here the migration base population is estimated according to NGQPOPY(0) + ½[BIR(0, t) − DEAY(0, t) + NETMOVY(0, t)].

2.2f *Final Adjustments to* CM II

The estimates of births, deaths, and elderly population are all scaled by factors λ_B, λ_D, and λ_E to sum to the respective national estimates (discussed in Part 1). These factors are constant over the 50 states and the District of Columbia. The estimates of young population for each state,

POPY(0) + λ_BBIR(0, t) − λ_DDEAY(0, t)

$\qquad$ + NGQMIGY(0, t) + GQPOPY(0, t) + NETMOVY(0, t), are then scaled to equal the estimate of national young population. The changes in the estimated young state population brought about by this last scaling are all attributed to the estimate of NGQMIGY(0, t). Estimates of the other components are not altered.

2.3 SOURCES OF DATA AND ERROR IN CM II

2.3a *Elderly Population:* POPE(t)

Certain characteristics of the Medicare data are pertinent here. Computer files prepared annually as of July 1 by the Health Care Finance Administration contain the state and county of the residential mailing addresses of persons enrolled in Medicare.

Problems arise in three areas: coverage, multiple addresses, and timing. At the national level, Medicare enrollment is generally about equal to the 1970 census count of the elderly (Bureau of the Census, 1973b). There is some disparity, however, between the census counts and the enrollment figures for some states, particularly Florida and Arizona. Various groups are excluded from Medicare (e.g., aliens who have resided in the country less than 5 years), and other groups are only partially included (e.g., retired federal employees are incompletely registered). A minor coverage problem arises from timing. Because of legislative requirements, the files as of a given date contain not only those aged at least 65 but also those who will turn 65 during the month following the reference date. Thus changes in Medicare enrollments reflect more closely the changes in the population over age 64-11/12 than those in the population 65 and over; the impact of this is minor, however.

The problem of multiple addresses occurs when an elderly person maintains residences in more than one state. Such a person may retain an original enrollment mailing address for Medicare purposes but by census definition be considered to be living at a residence in another state. Changes in the number of these persons would adversely affect the estimates of the states' elderly populations.

The timing problem derives from the delay in preparing the computer files. In order to be included in the Medicare files, a person must be 64-11/12 years old by July 1. The computer files are not updated and released, however, until about April 1 of the following calendar year, which is the closing date for new registrations or changes of address. Thus, for example, the actual reference date of the Medicare residence record for the July 1, 1976, elderly population is closer to April 1, 1977, than to the desired July 1, 1976. Changes in the number of persons misclassified would induce error into the estimates.

2.3b *Births, Young Deaths:* BIR(0, *t*), DEAY(0, *t*)

In making postcensal estimates of population prior to 1979, these two components of net natural increase for the young population were estimated on the basis of data provided by members of the Federal-State Cooperative Program (FSCP). The FSCP obtains its figures from the individual state vital statistics offices, and some error arises when national mortality rates by age and race are used to apportion the FSCP counts of total deaths in each state into estimates for the young and elderly populations. Under the revised procedures used to estimate 1979 populations (see section 2.2c) this source of error will be eliminated.

2.3c *Change in Group Quarters Young:* GQPOPY(0, *t*)

Two assumptions are used in estimating GQPOPY(0, *t*). The first is that except for the military barracks for which data are available, Job Corps centers, and refugee centers, there is no net migration of young group quarters residents into or out of the state. The error introduced by this assumption is believed to be small, since the change in number of out-of-state residents in college dorms is usually relatively small, as are changes in the number of residents living in barracks for which data are not available and other special populations whose changes are ignored (long-term inmates of hospitals and institutions).

The statewide change in the size of military barracks populations is estimated by summing the changes estimated for all subcounty barracks populations. For the large (and some small) military barracks an estimate of the size of the barracks at time *t* is secured from the individual post commander, either directly or through a member of the FSCP (see Bureau of the Census, 1973d, pp. 45-50). If these data are not available, data on the size of the whole installation are available, and the current barracks population is estimated to equal

$$\text{(current installation size)} \cdot \frac{(1970 \text{ barracks population})}{(1970 \text{ installation size})}.$$

This alternative was used for about 12 states in 1976–1978 and for five or six states in 1979.

The total change in group quarters young populations is thereby estimated to equal the movement in barracks populations, Job Corps centers, and refugee centers. This procedure utilizes the second assumption: the number of deaths in this subpopulation is zero.

2.3d *Non-Group Quarters Young Migration:* NGQMIGY$(0, y)$

The possible sources of error here are described by reference to the steps in the procedure outlined in section 2.2e.

Step I. School enrollment data are provided by members of the FSCP and by state education departments, based on figures to be supplied by grade by county for public and nonpublic schools. The roughly one-tenth of the states who do not have public school fall enrollment data available use year-end data.

Nonpublic school enrollments are reported (1) in some states by grade or by county by grade, (2) in some states for total kindergarten through grade 8 (here the Census Bureau tries to subtract kindergarten enrollment; these data, published in education directories, are not very accurate), and (3) in some states (such as Texas) only for some areas (in this case the parochial and private schools must be contacted in order to obtain enrollment figures; often the parochial school data can be obtained from a single source, but other private schools must be contacted one by one by an FSCP member or other means). Even when the states report private school enrollment, the Census Bureau screens the data.

Step II. Surprisingly, the estimate of ENROL(t) often exceeds that of SCLPOP(t). The reasons include the following: (1) children have been undercounted in the decennial census, (2) some children fail grades and are too old to be included in the estimate of SCLPOP(t), (3) students enrolled in special programs may be counted more than once, and (4) children of migrants and children who transfer from one school to another and are reported in both places are double-counted.

Step VI. The 1970 census included a question about prior residence in 1965. These data were used in estimating the number of migrants over the 5-year period for young females and for the school-age cohort.

2.4 ERROR STRUCTURE IN CM II

The principal source of error in the estimates of postcensal change in state populations resides in estimation of non-group quarters young migration NGQMIGY$(0, t)$. Such error arises because (1) misreporting (or nonreporting) of school enrollments introduces error into the estimates of the proportions enrolled in school, (2) differential undercoverage in the decennial census of the population under 14 adversely affects the estimates of the proportions enrolled in school, introducing error into the estimate of SCLMIGRAT(t) (see Step V above), and (3) the assumption and estimation of an unvarying linear change between the migration rates for school-age population and for young females are only rough approximations.

For a few states, notably Florida and Arizona, another significant source of error lies in the estimation of the change in the elderly population POPE(t) − POPE(0). Errors arise from deficiencies in the Medicare data (see section 2.3a above).

Error in the estimates of young deaths DEAY(0, t) is caused primarily by age and residence misreporting on death certificates. A smaller source of error lies in the adjustment to the national total by λ_D.

Errors in the estimates of births, caused by underregistration and misreporting of residence, are believed to be insignificant. Errors in the estimate of group quarters young migration are also believed to be generally insignificant, since in most states a very small proportion of the population lives in group quarters.

2.5 METHODOLOGY FOR RC

2.5a *Introduction and Notation*

Ratio-correlation (RC) is a regression method, in which a state population is divided into three parts: elderly, group quarters young, and non-group quarters young. The elderly and group quarters young populations are estimated as in CM II. In the case of non-group quarters young populations, RC uses regression equations to estimate the fraction of national non-group quarters young residing in each state. This fraction is then multiplied by the estimate of national non-group quarters young population, yielding an estimate of state non-group quarters young.

2.5b *Elderly Population*

The elderly component is estimated just as in the component method II (see section 2.2b above).

2.5c *Group Quarters Young Population*

The RC estimate of group quarters young population in the base year is obtained from the census count for April 1, 1970. To this is added an estimate of the change in group quarters young population (both in barracks and in nonmilitary group quarters), which is derived just as in the component method II (see section 2.2d). Deaths to group quarters young are ignored.

2.5d *Non-Group Quarters Young Population*

This component is estimated with the use of a regression equation. The equation is obtained by the least-squares linear fit of the relative changes in the state shares of national non-group quarters young population from 1960 to 1970 to the relative 1960–1970 changes in state shares of national numbers of (1) students enrolled in elementary school, (2) federal individual income tax returns, (3) registered passenger cars,[7] and (4) persons in the work force. The regression model has the form

$$Y_i = B_0 + \sum_{r=1}^{4} B_r X_{r,i} + \text{residual},$$

where B_0, B_1, ..., B_4 are the coefficients (to be estimated), Y_i is calculated by

$$\frac{\text{NGQPOPY}(t; i)/\sum_j \text{NGQPOPY}(t; j)}{\text{NGQPOPY}(0; i)/\sum_j \text{NGQPOPY}(0; j)}$$

with NGQPOPY(T; i) equal to non-group quarters population of state i at time T and $X_{r,i}$ having forms similar to Y_i but with NGQPOPY replaced by the predictor variables: observed numbers of students enrolled in elementary school, federal income tax returns, etc.[8]

The postcensal estimates of state non-group quarters young populations, for time t later than April 1, 1970, are obtained by using the estimated regression equation from above and substituting for the predictor variables the relative changes in shares of four components —students, tax returns, cars, work force—over the interval $(0, t]$. This yields an estimate of the relative changes in the state shares of non-group quarters young population. For each state this estimate is multiplied by the April 1, 1970, share of non-group quarters young population, to provide an estimate of the state's share of the national non-group quarters young population for time t. These estimates are then extrapolated 3 months to pertain to July 1 and scaled to sum to unity. Finally, these estimates are multiplied by the estimate of the national non-group quarters young population.

---

[7] This data series was dropped from use beginning with the 1975 estimates.

[8] Actually, the equation was developed in a far more complicated manner. These complications will be discussed in section 2.5f. For the time being, it is convenient to assume that the regression has been fitted as described above.

2.5e *Total Resident Population*

The total resident population for a state can now be estimated by adding the estimate of non-group quarters young population to the estimates of group quarters young population and of elderly population.

2.5f *Complications in Regression Models*

For this section the term "population" will be used to refer only to the non-group quarters young population. The regression complications relate to observed departures from the model of the predictor variables for some states. In particular, in almost every southern state the changes from 1960 to 1970 in the distribution of federal income tax returns, passenger car registrations, and to a lesser extent, the work force reflect increased affluence rather than changes in the state share of population only. Thus the deviations of regression-estimated non-group quarters young April 1, 1970, population from censal population counts have large positive values for the southern states. This same phenomenon was observed for the 1950–1960 changes.

For symptomatic data $V(T)$ referring to date T, the methodology focusses on "area coverage ratios," defined for state i as

$$R_i(T) = \frac{V_i(T)/P_i(T)}{V(T)/P(T)},$$

with notation

$R_i(T)$ area coverage ratio for variable V, state i, time T;
$V_i(T)$ value of variable V for state i, time T;
$P_i(T)$ population of state i, time T;
$P(T)$ $\sum_j P_j(T)$;
$V(T)$ $\sum_j V_j(T)$.

To improve the regression model, it is worthwhile to remove the effect of trends in the area coverage ratios. The "expected coverage ratio" for 1970, $R_i(70)$, is then calculated as follows:

1. If $R_i(50) < R_i(60) < 1$, then $R_i{'}(70)$ is established by linear extrapolation of the 1950–1960 trend, with a value of 1 as the upper limit; i.e., $R_i{'}(70) = \min[1, 2R_i(60) - R_i(50)]$.
2. If $R_i(50) > R_i(60) > 1$, then $R_i{'}(70)$ is established by linear ex-

trapolation of the 1950–1960 trend, with a value of 1 as the lower limit; i.e., $R_i{}'(70) = \max [1, 2R_i(60) - R_i(50)]$.

3. Otherwise, set $R_i{}'(70) = R_i(60)$.

If the trends in area coverage ratios are not being considered, then the predictor variable appearing in the regression equation for estimating 1960–1970 population change will be

$$\frac{V_i(70)/V_i(60)}{V(70)/V(60)}.$$

To account for the trends this variable is replaced by

$$\frac{V_i{}'(70)/V_i(60)}{V'(70)/V(60)},$$

where $V_i{}'(70) = V_i(70) \cdot R_i(60)/R_i{}'(70)$ and $V'(70) = \Sigma V_j{}'(70)$.

This replacement is in fact made for variables 2, 3, and 4 (section 2.5d) when the regression coefficients are estimated. To apply the estimated regression equation for estimation of postcensal population at time t later than April 1, 1970, each of the symptomatic variables corresponding to

$$\frac{V_i(t)/V_i(70)}{V(t)/V(70)},$$

is replaced by

$$\frac{V_i{}'(t)/V_i{}'(70)}{V'(t)/V'(70)},$$

where $V'(t) = \Sigma V_j{}'(t)$ and $V_j{}'(t)$ is calculated as follows. First the "expected area coverage ratio" $R_i{}'(80)$ is calculated analogously to $R_i{}'(70)$: if $R_i(60) < R_i(70) < 1$, then $R_i{}'(80) = \min [1, 2R_i(70) - R_i(60)]$, etc. Then $R_i{}'(t)$ is calculated by linear interpolation between $R_i(70)$ and $R_i{}'(80)$, and $V_i{}'(t) = V_i(t) \cdot R_i(70)/R_i{}'(t)$.

This use of area coverage ratios has been applied only for variables 2–4. Discussion can be found in the work of the Bureau of the Census (1974, pp. 10–14).

2.6 SOURCES OF DATA AND ERROR IN RC

2.6a *Elderly Population*

See section 2.3a.

2.6b Group Quarters Young Population

See section 2.3c.

2.6c Non-Group Quarters Young Population

Data on school enrollments were discussed in section 2.3d.

Information on individual income tax returns is made available to the Census Bureau by the Internal Revenue Service.

Data on passenger automobile registration are provided by the State Departments of Motor Vehicles and published by the Bureau of Public Roads in *Highway Statistics.*

Data on the numbers of nonagricultural wage and salary workers are provided by the U.S. Department of Labor and published annually in the May issue of *Employment and Earnings* (see Bureau of Labor Statistics, 1978, p. 158, pp. 124–133). Estimates of the number of full-time agricultural workers are based on data provided by FSCP members. Unemployment figures are currently obtained from the Bureau of Labor Statistics, which bases its figures on unemployment insurance data.

2.7 ERROR STRUCTURE IN RC

Most of the error in RC estimates of change in state populations derives from estimation of change in non-group quarters young population. This error arises in turn from error in the symptomatic data and from inadequacy of the regression model. Specifically, the model may fit well for a previous time period but predict poorly over the postcensal time period. The methodology discussed in section 2.5f applies only to known *past* departures from the model and not to *current* departures.

The comments in section 2.4 about error in estimating elderly and group quarters populations apply here to error structure of RC as well.

2.8 METHODOLOGY FOR AR

2.8a Introduction and Overview

The administrative records method (AR) is a relatively new variation of the component method for making postcensal population estimates. The components of population change are derived analogously with component method II (CM II), except for net migration. The elderly and special (group quarters) populations are handled separately, and natural increase is estimated identically. Net migration, however, is decomposed into net internal migration and immigration from abroad. To estimate net internal migration, individual federal income tax returns are matched for different years, and address changes noted. Immigration from abroad is estimated from the records of the Immigration and Naturalization Service on new immigrants' intended places of residence.

Another difference between AR and CM II lies in the "base year" used to estimate change. While CM II always calculates changes since the previous census, AR calculates shorter (usually year-to-year) changes. The objective in looking at shorter changes is the effort to obtain high match rates for the income tax returns. This will be explained further in the following sections.

2.8b Net Internal Migration

This component is estimated by computing a net migration rate for each state, based on state of residence reported on individual federal income tax returns for 2 years, and then applying this rate to the estimated young non-group quarters population. To develop 1973 postcensal estimates, the migration rate from 1970 to 1973 was estimated from the matching of calendar years 1969 and 1972 tax returns. The 1974 estimates were based on returns filed in April 1973 and 1974, and the 1975 estimates were based on returns filed in April 1973 and 1975. The 1976, 1977, and 1978 estimates were based on returns filed in April 1975 and 1976, 1976 and 1977, and 1977 and 1978, respectively.

The tax returns contain, for each filer, social security number, address, number of exemptions, and number of exemptions for blindness and/or old age.

For each of the calendar years when the tax forms were used, a computer file was constructed to retain the relevant information from the tax returns. The returns were arranged in ascending order of the social security number of the primary taxpayer.

No match is possible when the social security number on one year's return does not appear in the file for the other year. Reasons for this include the following: death; marriage; failure to earn sufficient income to require filing; immigration from abroad; first entry into the job market; divorce, separation, or widowhood (which may result in filing under a new social security number); and decisions by spouses to file jointly in one year but separately in another. A valid match can only occur if the social security number of the primary filer appears in both files. When the state of residence[9] is the same for both years, the filer (and any person claimed

[9] A question about state of residence appeared in the 1972 and 1975 returns. For other years, imputation procedures utilizing the mailing address on the return are used for estimating state of residence.

as an exemption) is classified as a nonmover. When the state of residence differs, the filer (and any person claimed as an exemption) is classified as an interstate migrant.

Because the elderly population is handled separately in AR, it is advantageous to exclude the elderly from the calculation of the non-group quarters young migration rate. Consequently, if any exemption is claimed for old age or blindness (the two are not distinguishable in the computer file), the entire tax return is excluded from consideration (i.e., treated equivalently to a nonmatched return).

On the basis of the remaining matched returns the migration rate is computed as

$$\frac{\left(\begin{array}{c}\text{number of exemptions on}\\ \text{inmigration returns}\end{array}\right) - \left(\begin{array}{c}\text{number of exemptions on out-}\\ \text{migration returns}\end{array}\right)}{\left(\begin{array}{c}\text{number of exemptions}\\ \text{on nonmover returns}\end{array}\right) + \left(\begin{array}{c}\text{number of exemptions on}\\ \text{outmigration returns}\end{array}\right)},$$

where "number of exemptions" refers to the tax return for the later of the 2 years. Except for minor complications (discussed in the following paragraph), this rate is multiplied by a population base equal to the number of young persons at the beginning of the period plus one-half the sum of natural increase plus net movement plus net immigration from abroad over the period. This product is the estimate of net non-group quarters young internal migration.

The possible complications in thus calculating the migration rate have been described by the Bureau of the Census (1976, p. 12) as follows:

Since migration patterns of young adults often differ from the remainder of the population, a migration adjustment factor distinct for each State was introduced. The rationale for the adjustment is that young adults are not represented on matched returns in proportion to their population. Accordingly, by reasoning analogous to that previously discussed in Component Method II, the net migration rate for the 10-year period 1960–70 was calculated for females under age 65 in 1970 and was compared to that of the subgroup which excluded those 18 to 24 in 1970. The algebraic difference between the two rates was the 10-year adjustment. For shorter periods the migration adjustment differential was prorated. At the State level, the annual adjustments range from −0.2 percent for West Virginia to +0.2 percent for Utah. The District of Columbia, however, receives an annual adjustment of +0.6 percent.

2.8c *Immigration From Abroad*

Immigrants from abroad are not detectable by the matching technique because they file tax returns only after entering the United States. The estimated national number of immigrants is allocated to states according to the immigrants' declarations to the Immigration and Naturalization Service. Emigrants are ignored. Parolees (see section 1.2c) receive special treatment.

2.8d *Other Components of Change*

These components include natural increase of young, changes in elderly populations, and changes in group quarters populations and are estimated as they are in CM II. While CM II considers changes over intervals $(0, t_1]$, $(0, t_2]$ (see section 2.2), AR focuses on $(t_1, t_2]$. To estimate change over the interval $(t_1, t_2]$, AR simply uses the difference between the CM II estimates of change over $(0, t_1]$ and $(0, t_2]$.[10]

The change in state population is then estimated by summing natural increase, changes in group quarters populations, changes in the elderly population, net internal migration, and immigration from abroad. The estimates of change in state population are scaled so their sum equals the change in the estimates of national population. As with CM II (see section 2.2f), the changes in the estimated young state population brought about by this last scaling are all attributed to the estimate of net internal migration of the non-group quarters young.

The postcensal estimates of state population are then obtained by adding these estimates of population change to the estimates of population in the base year.

2.9 SOURCES AND STRUCTURE OF ERROR FOR AR

Since postcensal estimation of state population under AR differs from CM II with respect to the migration component only, the focus here will be on the use of individual federal income tax returns to estimate migration. The methodology rests on two implicit assumptions:

1. Migration patterns are the same for people who file income tax returns as for those who do not (except for elderly and special populations, which are treated separately).

2. The address listed on the tax form for each year is the filer's residence and is the relevant address for determining whether or not the

---

[10]Beginning with the 1978 estimates, the Census Bureau computed deaths to the young and to the elderly over $(t_1, t_2]$ directly rather than by taking differences between those over $(0, t_1]$ and $(0, t_2]$. The two procedures are not equivalent because the cohorts of young (and elderly) at times t_1 and t_2 were different, and what is really of interest are deaths over $(t_1, t_2]$ to the cohort defined with t_2 as the reference date.

person has moved. For example, the filer could report place of residence one year and place of business in another year.

The extent of error arising from failure of assumption 1 is not known. About 99 percent of whites under 65 are included as exemptions on the tax returns, but the filing rate for blacks under 65 is lower. Blacks in the southern states have exceptionally low filing rates. Also, numerous low-income persons are not included as exemptions, when the head of household fails to file a tax return. Further discussion is found in the work of the Bureau of the Census (1978a, pp. 4, 6).

Estimates of net internal migration to New York are probably understated because the matching-based estimates of Puerto Rico–New York migration are based on underestimates of Puerto Rico–New York migration but more accurate estimates of New York–Puerto Rico migration. This arises from peculiarities in the tax laws as they affect Puerto Rico. Persons living in Puerto Rico typically do not need to file an Internal Revenue Service (IRS) individual income tax return and so will not be matched if and when they migrate to New York. On the other hand, most Puerto Ricans returning from New York to Puerto Rico will probably file a tax return with IRS (to recover withholding taxes), giving Puerto Rico as place of residence.

Errors in postcensal estimates of state populations connected with the estimates of net migration from abroad arise from (1) errors in allocating the immigrants to the correct states, (2) errors in the estimate of the total number of immigrants from abroad, and (3) treatment of emigrants by foreign countries. Discussion of point 2 as a source of error can be found in section 1.3 above. The effect of point 3 is complicated because of the adjustment of total migration to the national control.

PART 3   COUNTY ESTIMATES

3.1 INTRODUCTION

Postcensal estimates of county populations are calculated by methods generally similar to those discussed in Part 2. Other methods may also be utilized at the substate level because some states prepare their own estimates. These are scaled to sum to the Census Bureau's estimate of the state total and then averaged with the Bureau's estimates of substate populations.

It is important to distinguish among three sets of county estimates: "provisional," "preliminary" or "ORS" (for Office of Revenue Sharing), and "revised." Provisional estimates are made roughly 6–12 months after the reference date for the estimates, the revised estimates about a year later, and the ORS estimates sometime in between.

Because the provisional estimates are made before the Internal Revenue Service (IRS) tax return data are available, these estimates do not employ the administrative records method (AR). Rather, component method II (CM II) is used to estimate the population change over the year preceding the estimate date t, by calculating the difference between the CM II estimates for t and $t - 1$. In the case of large metropolitan counties the housing unit method is generally also used to estimate the 1-year population change. For these counties the estimates of change from the housing unit method and CM II are averaged. The derivation of the provisional county estimates may be represented symbolically as provisional estimate $(t) =$ revised estimate $(t - 1)$ + change over $(t - 1, t]$, where change over $(t - 1, t]$ is estimated either by the change in CM II estimates alone from date $t - 1$ to date t or by the average of the changes in CM II and the housing unit method estimates from $t - 1$ to t. In several states (18 for the 1977 estimates and 16 for 1976) other methods supplant the housing unit method in computing the provisional estimates.

Generally, the ORS estimates are derived according to

$$\text{ORS estimate } (t) = \text{revised estimate } (t - 1) + \text{change over } (t - 1, t],$$

where change over $(t - 1, t]$ is estimated by the average of change in CM II and AR estimates from $t - 1$ to t. In some states, additional methods are averaged to estimate change over $(t - 1, t]$. However, the Census Bureau requires that estimates within a state be the product of a uniform methodology, so additional methods are averaged only if they provide estimates for all counties in a state. Thus for the 1975 ORS estimates, the housing unit method was used in only one state (Florida), where the housing unit method estimates were available for all counties without exception.

When the results of a special census are available for a county, they are used instead of the various postcensal estimates. In this case the adjustment of county estimates to sum to the state estimate follows a complicated procedure, which we will refer to here as "rake/float." This procedure is discussed below in section 4.2 for subcounty estimates. The procedure for county estimates is analogous and will not be explicitly given.

The notation and conventions introduced in Part 2 will be retained in the present and subsequent chapters.

For the July 1, 1975, ORS county estimates for Florida, the change was estimated by a three-way average of changes in AR, CM II, and housing unit method estimates. (For other exceptions, see Bureau of the Census (1980).) For Kansas, Missouri, Nebraska, and Washington July 1, 1975, ORS county estimates, the change was estimated by the average changes in CM II, AR, and RC estimates. For California, four estimates of change were averaged: CM II, AR, RC, and the driver's license address change method (DLAC).

Revised estimates of county population differ in structure from both provisional and ORS estimates. The procedure in making the revised estimate for date t does not employ the revised estimate for date $t - 1$ explicitly. Rather, CM II, AR, and RC are each used directly to estimate the population as of date t (in ways similar to those described in Part 2). In some states a fourth method is used as well. Each method's set of county estimates is scaled to sum to the state total, and then the three (or four) estimates for each county are averaged. This procedure yields the revised estimates of county populations.

The various methods are described below.

3.2 DRIVER'S LICENSE ADDRESS CHANGE METHOD

The driver's license address change method (DLAC) is a component method used by California to estimate county populations. The estimates are constructed in the following manner: to the base population estimates are added estimates of natural increase, plus change in the elderly populations (estimated from changes in Medicare enrollments), plus changes in military barracks, plus net migration. The distinguishing feature of DLAC is the way in which net migration is estimated.

Net interstate migration of the population aged 18 to 64 is estimated using address changes in the California Driver's License File. Persons outside this age range are not well represented, and their migration is estimated separately. Immigration from abroad is also estimated separately. A variation of CM II is used to estimate net migration of the population under 18. Migration of persons over 64 is implicitly included in the estimate of changes in the elderly population. Further detail is given by Rasmussen (1975).

3.3 HOUSING UNIT METHOD

The state-prepared county population estimates in Florida for 1975 were based on the housing unit method (HUM). In this method an estimate of the number of occupied housing units is made and multiplied by an estimate of the average number of persons per household. To this product is added an estimate of special populations not in housing units, yielding an estimate of total population. Further discussion appears in the work of Starsinic and Zitter (1968) and Pittenger et al. (1977).

3.4 COMPONENT METHOD II

The use of CM II for counties essentially parallels that for states (see section 2.2 above), with exceptions noted in the following descriptions of the methods used in connection with each component.

3.4a *Elderly Population*

This component is estimated just as at the state level (see section 2.2b above).

3.4b *Special Populations*

Because group quarters populations may account for a more significant share of a county's population than of a state's population, these populations are estimated more painstakingly at the county than at the state level. The group quarters populations considered at the county level include inmates of prisons or of long-term hospitals, college students living in dormitories, residents of Job Corps centers, and members of the armed forces living in military barracks.

For these special populations, annual observations are obtained and net changes over the year are computed. Net movement of the barracks populations for counties is estimated by allocating the state total among the counties, according to the 1970 census distribution of males aged 14–17.

3.4c *Births and Deaths to Young*

These components are estimated analogously to their state-level counterparts, with two major differences. The first difference is that at the county level the Census Bureau does not use reported county deaths by race. The second difference is that the reported births and deaths for the counties are not adjusted to the state total (which had been adjusted to the national total). Thus births are estimated simply by obtaining the number of reported births for each county from state vital statistics departments through members of the Federal-State Cooperative Program (FSCP).

Young deaths for counties over the interval $(0, t]$ are estimated as follows: Let subscripts r, a, i, and j refer to race, age, state, and county and let the argument x refer to the year ending December 31. Race r takes on two values (white, black and other), as does age a (young, elderly). Define

$D_{ij}(x)$ reported number of deaths for county j, state i, year x (obtained from state vital statistics departments);

d_{ra} NCHS estimate of the nationwide period death rate for persons of race r and age group a over the interval $(0, t]$;

P_{raij} count of race r, age a population of county j in state i on April 1, 1970.

The estimated number of deaths for county j, state i over the interval $(0, t]$ will be denoted by D_{ij} and is obtained by summing $D_{ij}(x)$ over time periods x and interpolating at the ends of the interval. For example, with t referring to July 1, 1973, D_{ij} satisfies

$$D_{ij} = .75 D_{ij}(70) + D_{ij}(71) + D_{ij}(72) + .5 D_{ij}(73).$$

The sum of D_{ij} over counties j is not controlled to a state total.

An initial estimate, D'_{aij}, of the number of deaths over $(0, t]$ to age cohort a in county j, state i is obtained by applying the national period death rates by age and race to the corresponding county cohorts in 1970 and summing over races:

$$D'_{aij} = \sum_r P_{raij} d_{ra}.$$

These initial estimates are then used to apportion the reported county deaths into those for the two age groups. Thus the deaths to the young in county j, state i over the interval $(0, t]$ are estimated by

$$\text{DEAY}(0, t; i, j) = D_{ij} \frac{D'_{yij}}{D'_{yij} + D'_{eij}}$$

where a takes on the values y (young) and e (elderly).

3.4d Non-Group Quarters Migration of the Young

This component is estimated essentially as at the state level (see section 2.2e) with certain differences. For counties the base period school-age and young female migration rates SCLMIGRAT(0) and FEMIGYRAT(0) are 10-year rates, calculated over the previous intercensal decade. In addition, the school-age and young female migration rates for the base period and for the current postcensal period are each multiplied by factors to account for underexposure of the entire cohort to migration. For example, in calculating the base period migration rate for the school-age population (aged 6.25 to 14.24 on April 1, 1970), allowance is made for the fact that children aged 6.25 to 7.25 on April 1, 1970, were only exposed to migration (on the average) 6.75 years rather than the full 10 years. (More details are given by van der Vate (1978).) Thus the analog for counties of step VII in section 2.2e calculates

$$\text{FEMIGYRAT}(t) = \text{SCLMIGRAT}(t) + [\text{FEMIGYRAT}(0)$$
$$- \text{SCLMIGYRAT}(0)](t/10),$$

where the various rates have been multiplied by the factors for underexposure.

3.4e *Adjustment to Totals*

The process of adjusting to totals is the same at the county as at the state level, except that births and deaths are not adjusted separately. Thus the factors λ_B and λ_D, as stated in section 2.2f above, are both set equal to unity.

3.5 SOURCES OF DATA AND ERROR IN CM II

The discussion in section 2.3 above applies here as well. In addition, problems with Medicare data and group quarters migration estimates become more severe at the county level. Some counties (especially in Florida) do not have complete Medicare coverage (see Irwin, 1978). Furthermore, differential coverage of the elderly population by Medicare has more significant impact for counties than for states.

Geographic coding of Medicare records is also problematic, in that address codes are derived largely from the names of cities, some of which spread across county lines. In addition, extensive areas beyond the limits of a city frequently carry the city name. When such areas extend into a second county, the addresses are apt to be coded to the county containing the major part of the city. The independent cities in Virginia especially are affected in this way, so that estimates of the elderly populations of the adjoining counties are subject to large error (see Irwin, 1978, pp. 13-15). Another source of error arises when a Medicare enrollee who has never filed for benefits makes an address change for social security purposes—in particular if the person's social security check is mailed directly to a bank—and the Medicare address is automatically changed to agree with the social security address.

The data on group quarters populations are provided by state agencies involved in the Federal-State Cooperative Program (FSCP). The county figures are sums of the figures for subcounty areas (see section 4.1f for more discussion).

3.6 STRUCTURE OF ERROR IN CM II

The error structure in CM II at the county level roughly parallels that at the state level, except that the components of error are larger at the county level. (See section 2.4 above for relevant discussion.)

3.7 ADMINISTRATIVE RECORDS METHOD

Postcensal estimation using AR is approximately the same for counties as for states. All components except net migration are estimated just as in CM II (see sections 3.4, 3.5, and 3.6 above).

Data on place of intended residence for resident aliens (immigrants who declare their intentions to secure U.S. citizenship) are kept by the U.S. Immigration and Naturalization Service for states and places with 1970 populations of 100,000 or more. Explicit estimates of the number of immigrants from abroad are made for areas within a state having fewer than 100,000 residents in 1970 by the use of the number of persons of foreign birth reported in the 1970 census. Estimates of immigration from abroad to counties are derived by summing the estimates of immigration to places within the county.

Estimation of the county-level, young non-group quarters net migration resembles that for the state level. Let s and t denote the time references for the base population and the current estimate, respectively. The young non-group quarters net migration rate IRSRAT($s, t; i, j$) for county j in state i is calculated as

$$\text{IRSRAT}(s, t; i, j) = \frac{\text{INS}(s, t; i, j) - \text{OUTS}(s, t; i, j)}{\text{OUTS}(s, t; i, j) + \text{NONMOV}(s, t; i, j)},$$

where INS($s, t; i, j$) is the number of exemptions on matched individual federal income tax returns classified as inmigrants to county j in state i over the period $(s, t]$, such that the tax returns did not have exemptions for age or blindness; OUTS($s, t; i, j$) is the number of exemptions on matched individual federal income tax returns classified as outmigrants from county j in state i over the period $(s, t]$, such that the tax returns did not have exemptions for age or blindness; and NONMOV(s, t; i, j) is the exemptions on matched individual federal income tax returns classified as nonmovers from county j in state i over the period $(s, t]$, such that the tax returns did not have exemptions for age or blindness.[11]

Exemptions on a matched return are classified as inmigrants (out-migrants) if the designated county for the address on the return differs at dates s and t and the address at date s (date t) is in county j in state i (see Galdi, 1978). An exemption on a matched return is classified as a non-mover if the address on the return is designated to be the same in county j in state i for both dates s and t. The number of exemptions refers to the number at date t.

To estimate the net migration for the young non-group quarters population, the migration rate IRSRAT is multiplied by a population base MIGBASE defined by

$$\text{MIGBASE}(s, t; i, j) = \text{NGQPOP}(s; i, j)$$

$$+ \frac{1}{2} \{ \text{BIR}(s, t; i, j) - \text{DEAY}(s, t; i, j)$$

$$+ \text{NETMOVY}(s, t; i, j) + \text{IMM}(s, t; i, j) \},$$

with notation

NGQPOP(s; i, j)   young non-group quarters population at date s in county j in state i;

NETMOVY(s, t; i, j)   net movement of young from military population overseas to resident civilian population in county j in state i over the period $(s, t]$;

IMM(s, t; i, j)   immigration from abroad to county j in state i over period $(s, t]$.

3.8 ADJUSTING ADMINISTRATIVE RECORDS METHOD ESTIMATES TO TOTALS

As in the methods discussed above, AR estimates of county populations are scaled to sum to the estimate of state population. Changes in the county estimates effected by this last scaling are all attributed to the net migration component.

---

[11] These exclusions were made because the elderly population is estimated separately (see section 2.8b for further discussion).

3.9 SOURCES OF ERROR IN THE ADMINISTRATIVE RECORDS METHOD

At the county level, residence classification is difficult because a mailing address is not always sufficient to determine county of residence. A major problem arises when the post office in a city of one county serves residents of an adjacent county; thus people report their addresses as being in the city of the post office rather than in that of their residence. This problem typically occurs when a town straddles county boundaries or when adjacent counties have towns with the same name. To ameliorate this problem, a special question was placed on the 1972 and 1975 tax forms to obtain information on state, county, incorporated place, and township of residence. Galdi (1978) has described the use of the data obtained from this question.

The discussion of state-level error for AR in section 2.9 above applies to the county level as well.

3.10 RATIO-CORRELATION METHOD

Estimation of county populations using RC differs somewhat from state-level estimation: at the county level the elderly population is not treated separately. Thus at the county level, RC estimates total non-group quarters population rather than non-group quarters young population. Otherwise, the estimation of non-group quarters population is the same as for states except that (1) the kinds of symptomatic data used vary for different states (see Bureau of the Census (1980) for details) and (2) the complications involving "area coverage ratios" (discussed toward the end of section 2.5f above) are not introduced at the county level.

The discussion of RC in sections 2.5, 2.6, and 2.7 above is thus relevant here as well.

3.11 USE OF SPECIAL CENSUSES

If a special census was taken for a county within a year of the postcensal estimate date, the special census count replaces the average of the methods' estimates for that county. Since special censuses usually do not fall precisely on July 1, the counts are typically interpolated backward or extrapolated forward according to the trend since April 1, 1970.

Using the results of the special census for succeeding updates is straightforward for the AR method, which estimates population change since the last update. The special census count is reflected in the estimate of BASEPOP. Component method II and the ratio-correlation method, however, always refer to changes since the last decennial census, which makes it more difficult to use past special censuses in succeeding updates when these methods are used.

For illustrative purposes, suppose a special census were conducted on July 1, 1975, and that the CM II and RC estimates for this date were 1,200 and 1,000 lower, respectively, than the special census count. For the 1975 estimate the special census count would be used. For the 1977 estimate the 1975 special census would be reflected in the BASEPOP estimate used by AR, but it would not be reflected in the CM II and RC estimates. The Census Bureau would make use of the 1975 special census by adding 1,200 and 1,000 to the 1977 CM II and RC estimates. The implicit assumption is that either (1) the methods are biased (i.e., the assumptions don't apply to the county under consideration), (2) the 1970 data are in error, or (3) the 1975 data are lagging in indicating population change. According to point 1 or 2 the CM II and RC estimates would be too low throughout the decade. According to point 3 the CM II and RC estimates would be too low for a while but would ultimately catch up to the true level of change. If point 3 were relevant and the CM II and RC estimates did catch up, the Census Bureau would like to stop adding 1,200 and 1,000 to the CM II and RC estimate. To determine whether the estimates were catching up, the Census Bureau would monitor the time series of population changes as estimated by the different methods and look for sharp shifts occurring in RC or CM II but not in AR. If this were noted, the Census Bureau would stop adding in the differences between the special census and the method's estimates, 1,200 and 1,000, and no further explicit consideration of the post special census would be taken.

PART 4   SUBCOUNTY ESTIMATES

The administrative records method (AR) is generally the only method used to make postcensal population estimates for subcounty units. However, the results of recent special censuses are used when available, in lieu of the AR estimates. When the special census estimates are used, the adjustment of the subcounty estimates to sum to county estimates follows a complicated procedure, sometimes called "rake/float," to be discussed below.

A few states provide subcounty estimates of their own. These are scaled to sum to the county estimates and then averaged with the AR estimates.

4.1 ADMINISTRATIVE RECORDS METHOD

Let time $T = 0$ refer to April 1, 1970, and let T be scaled in years. The time references for the AR estimates, s and t, will correspond to the time references for the base population and the current estimate, respectively. The notation introduced below will refer to subcounty unit k in county j of state i.

The resident population estimate $\text{AR}(t; i, j, k)$ consists of seven elements as follows:

non-group quarters population at time s $(\text{NGQPOP}(s; i, j, k))$
   $+$ births to residents over the interval $(s, t]$ $(\text{BIR}(s, t; i, j, k))$
   $-$ deaths to residents over the interval $(s, t]$ $(\text{DEA}(s, t; i, j, k))$
   $+$ net non-group quarters inmigration over the interval $(s, t]$
      $(\text{NETMIG}(s, t; i, j, k))$
   $+$ immigration from abroad over the interval $(s, t]$ $(\text{IMM}(s, t; i, j, k))$
   $+$ population in military barracks at time t $(\text{MILBAR}(t; i, j, k))$
   $+$ members at time t of special populations other than military barracks residents $(\text{IC}(t; i, j, k))$.

It is important to notice that the elderly population is no longer treated separately, because Medicare data are not available for measuring change below the county level.

Each of the above elements will now be discussed in turn.

4.1a *Non-Group Quarters Population at Time s:* $\text{NGQPOP}(s; i, j, k)$

Let $\text{ORS}(s; i, j, k)$ denote the final estimate of resident population for date s. The notation "ORS" is appropriate because this estimate of population is used by the Office of Revenue Sharing. Then NGQPOP is calculated by

$$\text{NGQPOP}(s; i, j, k) = \text{ORS}(s; i, j, k) - \text{MILBAR}(s; i, j, k) - \text{IC}(s; i, j, k).$$

4.1b *Births Over (s, t]:* $\text{BIR}(s, t; i, j, k)$

Neither NCHS nor state vital statistics offices compile data on resident births for most of the places of population under 10,000 (more than half of the subcounty units).

In estimating $\text{BIR}(s, t; i, j, k)$ the following procedure is used to allocate reported county births to all the subcounty units for which reports of births are questionable or not available.

First, the area-specific, age-adjusted fertility rate for the census year 1970 is established, according to the distribution of the population under 1 year old on April 1, 1970. The proportion of population aged under 1 year in county (i, j) living in subcounty area (i, j, k) is calculated according to

$$\mathrm{PU}1(0; i, j, k) = \frac{\mathrm{U}1(0; i, j, k)}{\sum_k \mathrm{U}1(0; i, j, k)},$$

where $\mathrm{U}1(0; i, j, k)$ is the population aged under 1 on April 1, 1970, in area (i, j, k).

The number of births in the year ending April 1, 1970, is then estimated by

$$\mathrm{B}(0; i, j, k) = \mathrm{PU}1(0; i, j, k) \times \mathrm{B}(0; i, j),$$

where $\mathrm{B}(0; i, j)$ is the number of births to residents of county (i, j) during the calendar year 1970.

The fertility rate for women 15 to 39 years old in area (i, j, k) on April 1, 1970, is calculated by

$$\mathrm{FR}(0; i, j, k) = \frac{\mathrm{B}(0; i, j, k)}{\mathrm{F}1539(0; i, j, k)},$$

where $\mathrm{F}1539(0; i, j, k)$ is the number of women aged 15 to 39 on April 1, 1970, residing in non-group quarters.

This census year fertility rate is then applied to estimate births during the following year:

$$\mathrm{B}(1; i, j, k) = \mathrm{FR}(0; i, j, k) \times \mathrm{F}1539(1; i, j, k),$$

where $\mathrm{F}1539(1; i, j, k)$ is the number of women aged 15 to 39 on April 1, 1971, residing in non-group quarters.

To estimate $\mathrm{F}1539(1; i, j, k)$, the following procedure is used. A net migration rate for the young population for the year $(0, 1]$ is estimated from matching of IRS tax returns. Essentially, this rate is calculated analogously to IRSRAT, described in section 3.7 above. Denoting this migration MIGYRAT$(1; i, j, k)$, calculate

$$\mathrm{F}1539(1; i, j, k) = \mathrm{F}1539(0; i, j, k)$$
$$+ \mathrm{F}1539(0; i, j, k) \times \mathrm{MIGYRAT}(1; i, j, k).$$

Recursively, for time $T = 2, 3, 4, \ldots, 9$, calculate

$$\mathrm{F}1539(T; i, j, k) = \mathrm{F}1539(T - 1; i, j, k)$$
$$+ \mathrm{F}1539(T - 1; i, j, k) \times \mathrm{MIGYRAT}(T; i, j, k),$$

where MIGYRAT(T; i, j, k) refers to migration over the interval $(T - 1, T]$.[12] For noninteger values of T, F1539(T; i, j, k) is computed by linear interpolation.

The annual resident births subsequent to 1970 are estimated on the basis of the female population 15–39, estimated as above. To maintain consistency with the annual birth statistics for the county resident population, however, these estimated resident births for area (i, j, k) are adjusted (ADJB) to sum to the county total B(T; i, j, k):

$$\text{ADJB}(T; i,j,k) = \text{B}(T; i,j,k) \times \frac{\text{B}(T; i,j)}{\sum_{k} \text{B}(T; i,j,k)}.$$

Further adjustments incorporated in ADJB will be discussed below. Note that we have yet to derive B(T; i, j, k) for $T > 1$.

To estimate B(2; i, j, k), we make use of ADJB(1; i, j, k) to update the fertility rate, so

$$\text{FR}(1; i,j,k) = \frac{\text{ADJB}(1; i, j, k)}{\text{F1539}(1; i, j, k)}$$

and

$$\text{B}(2; i, j, k) = \text{FR}(1; i, j, k) \times \text{F1539}(2; i, j, k).$$

For integer $T > 1$ the formulas are

$$\text{FR}(T; i,j,k) = \frac{\text{ADJB}(T; i, j, k)}{\text{F1539}(T; i, j, k)}$$

and

$$\text{B}(T; i, j, k) = \text{FR}(T - 1; i, j, k) \times \text{F1539}(T; i, j, k).$$

On the basis of the distribution of estimated births by place, "tolerance intervals" are constructed (see Cavanaugh, 1977, pp. 33–35). Recall that

---

[12] Specifically, MIGYRAT(T; i, j, k) is calculated as IRSRATY(s, T; i, j, k)/($T - s$) where s is the latest time prior to T for which the tax file is available and IRSRATY(s, T; i, j, k) is calculated the same way as IRSRAT(s, T; i, j, k) in section 4.1e, except that only returns not claiming exemptions for old age or blindness are used. (This is the same set of returns used to estimate county rates IRSRAT(s, T; i, j); see section 3.7 above.)

for roughly half of the subcounty units, information on reported births is available from NCHS or from the state vital statistics offices. These reported figures are accepted as estimates of births only if they fall within the appropriate tolerance interval. Otherwise, the reported figures are replaced by the estimates derived above. These estimates are also used if no reported data are available. At this point, the estimates of births for subcounty units are adjusted to sum to the estimate of total county births.

4.1c *Deaths Over (s, t]:* DEA(s, t; i, j, k)

As in the case of resident birth reports, information about deaths is not available from NCHS or state vital statistics offices. Thus for over half of the subcounty units, deaths must be estimated by indirect methods, rather than by direct reports of deaths. The procedure to estimate deaths applies effectively the same logic that underlies the estimation of births, described in section 4.1b above.

While the estimated female population aged 15–39 composes the basic reference for consideration of birth events, age distributions of the estimated populations of subcounty areas are the most direct referent in estimating deaths. Hence the "young" population (under 65 years old), "elderly" (over 65), and deaths occurring to these two broad age groups are treated separately in allocating county resident deaths to subcounty areas. Racial differences in mortality as well are handled in the subcounty estimation by an allocation of white and nonwhite deaths according to the racial distributions observed in the 1970 decennial census. After the annual estimates of young and elderly populations are separated into white and nonwhite components according to the 1970 proportionality, the allocation of resident deaths in the county among the subcounty areas proceeds similarly for both racial categories. Thus the following description of the allocation procedure will denote all nomenclature by w for the white population, and not repeat the same description for the nonwhite population.

First, the area-specific, age-race-specific death rates for calendar year 1970 are established. Members of the Federal-State Cooperative Program obtain counts of the total deaths in each county by contacting state vital statistics departments. These deaths are allocated to the four age-race groups (young and elderly by white and nonwhite) in the county on the basis of statewide death rates for the four groups. These death rates are estimated from life tables constructed by the National Center for Health Statistics. The county-level deaths for each of the four age-race groups are then prorated by age-race to each subcounty unit according to the unit's share of the county population. For example, let DEAYW(0; i, j) be the estimate of deaths to young whites of county j, state i in calendar year 1970 and let POPYW($0; i, j, k$) be the number of young white non-barracks residents of subcounty area k on April 1, 1970. The number of young white deaths in subcounty area k during the calendar year 1970 is estimated according to

$$\text{DEAYW}(0; i,j,k) = \text{DEAYW}(0; i,j) \times \frac{\text{POPYW}(0; i,j,k)}{\sum_k \text{POPYW}(0; i,j,k)}.$$

The corresponding estimation for the elderly population is, with corresponding notation,

$$\text{DEAEW}(0; i,j,k) = \text{DEAEW}(0; i,j) \times \frac{\text{POPEW}(0; i,j,k)}{\sum_k \text{POPEW}(0; i,j,k)}.$$

The death rates in 1970 are then calculated as

$$\text{DEAYWRAT}(0; i,j,k) = \frac{\text{DEAYW}(0; i,j,k)}{\text{POPYW}(0; i,j,k)}$$

and

$$\text{DEAEWRAT}(0; i,j,k) = \frac{\text{DEAEW}(0; i,j,k)}{\text{POPEW}(0; i,j,k)}.$$

These death rates are applied to the respective estimates of population for the subsequent year, 1971, for an estimate of resident deaths in that year. The annual estimates of resident deaths in subcounty areas are in turn adjusted to the county total. Again recursively, the adjusted area deaths are used to compute the area death rate, to be used for the estimate of resident deaths in the succeeding year. The procedure is thus quite similar to that for births. The annual estimates of the population by age-race for each subcounty area will now be described in some detail.

Since the component of population change by death is considered for the non-group quarters population only, the annual estimates of population used to multiply the death rates must be diminished by estimates of the group quarters population and some part of the net movement from non-group to group quarters population over the year. In practice, the military barracks population alone is considered rather than the entire group quarters population at time $T = 1$ (1 year after $T = 0$, the date of April 1, 1970). The non-barracks young population NBY($1; i, j, k$) is initially (before deaths) estimated by

NBY(1; i, j, k) = POPY(0; i, j, k) − MILBAR(0; i, j, k)

+ {POPY(0; i, j, k) − MILBAR(0; i, j, k)} × MIGYRAT(1; i, j, k),

where MILBAR refers to the barracks population (assumed all young) and MIGYRAT is the migration rate for young persons (described in section 4.1b above). On the basis of the racial composition observed in the 1970 census, the estimate NBY(1; i, j, k) is partitioned into estimates of the white and nonwhite subpopulations. These race estimates are then multiplied by the death rates computed earlier, yielding estimates of deaths by race to the young population.

The procedure for the elderly is analogous. The migration rate for the elderly, MIGERAT, is calculated in a manner similar to MIGYRAT, except that *only* tax returns with age or blindness exemptions are used. The elderly population at time $T = 1$ is initially (before deaths) estimated by

POPE(0; i, j, k) + POPE(0; i, j, k) × MIGERAT(1; i, j, k).

Then this estimate is partitioned into estimates by race (according to the racial composition observed in the 1970 census), and the death rates discussed above are applied to the respective initial estimates of the elderly population by race. This yields estimates of deaths by race to the elderly population.

For every subcounty unit in a county the estimates of deaths for each of the four age-race groups are separately scaled to sum to a county control. The scaled components are then added to yield an "adjusted" estimate of deaths over (0, 1] for the subcounty unit.

Tolerance intervals are constructed and used for deaths as for birth estimates.

To develop estimates of deaths for times later than $T = 1$, the procedure described above is applied recursively[13] in the manner outlined in section 4.1b for recursive estimation of births.

4.1d *Net Migration Over (s, t]:* NETMIG($s, t; i, j, k$)

The non-group quarters migration rate for subcounty units, IRSRAT($s, t; i, j, k$), is calculated analogously to IRSRAT($s, t; i, j$) as described for counties in section 3.7 above:

---

[13] For example, in deriving an initial estimate of the young or elderly populations for $T = 2$ (see two preceding displays for $T = 1$), allowance is made not only for migration but also for deaths over the interval (0, 1].

$$\text{IRSRAT}(s, t; i, j, k) = \frac{\text{INS}(s, t; i, j, k) - \text{OUTS}(s, t; i, j, k)}{\text{OUTS}(s, t; i, j, k) + \text{NONMOV}(s, t; i, j, k)},$$

where INS, OUTS, NONMOV for subcounty unit k are, with one difference, defined analogously to their county-level counterparts. The difference is that exemptions for age and blindness were excluded from the county analysis but included in the subcounty analysis. Thus at the county level, IRSRAT refers to the young only, but at the subcounty level it refers to both young and elderly. Thus we have

INS$(s, t; i, j, k)$ exemptions on matched individual federal income tax returns classified as inmigrants to subcounty unit k in county j, state i over the period $(s, t]$;

OUTS$(s, t; i, j, k)$ exemptions on matched individual federal income tax returns classified as outmigrants from subcounty unit k in county j, state i over the period $(s, t]$; and

NONMOV$(s, t; i, j, k)$ exemptions on matched individual federal income tax returns classified as nonmovers from subcounty unit k in county j, state i over the period $(s, t]$.

The returns are matched by social security number, and the number of exemptions refers to the number at date t.

4.1d(1) *The Special Problem of Residence Classification*   In making subcounty estimates an important procedural element involves the assignment of geographic locations to the tax returns. It should be noted that all problems concerning residence classification are greater at the subcounty than at the state or county level. In order to determine the governmental unit to which the exemptions on a given tax return should be referred, each tax return must be assigned a geographic code identifying the state, county, minor civil division if any, and city, borough, or village. Assignment of geographic codes is difficult because they cannot be accurately determined solely on the basis of mailing address (state and post office names) given by the filer of the tax return. For one thing, many subcounty governmental units do not have a post office. Moreover, the postal delivery area of a subcounty governmental unit that has a post office does not in general coincide with the unit's geographic boundaries. Finally, the mailing address and place of residence of the filer can differ.

176

Form 1040

US Department of the Treasury—Internal Revenue Service
Individual Income Tax Return **1975**

For the year January 1–December 31, 1975, or other taxable year beginning _____, 1975, ending _____, 19___

Please print or type	Name (If joint return, give first names and initials of both)	Last name	Your social security number	For Privacy Act Notification, see page 2 of Instructions.
	Present home address (Number and street, including apartment number, or rural route)		Spouse's social security no.	For IRS use only
	City, town or post office, State and ZIP code		Occu- Yours ▲ pation Spouse's ▲	

Requested by Census Bureau for Revenue Sharing

A In what city, town, village, etc., do you live? ▲

B Do you live within the legal limits of the city, town, etc.?
☐ Yes ☐ No ☐ Don't know

C In what county and State do you live?
County _____ State

D In what township do you live? (See page 4.)
☐

Filing Status **Exemptions**

1 ☐ Single (check only ONE box)

2 ☐ Married filing joint return (even if only one had income)

3 ☐ Married filing separately. If spouse is also filing give spouse's social security number in designated space above and enter full name here ▲

4 ☐ Unmarried Head of Household (See page 5 of Instructions)

5 ☐ Qualifying widow(er) with dependent child (Year spouse died ▲ 19___). See page 5 of Instructions.

6a ☐ Regular ☐ Yourself ☐ Spouse   Enter number of boxes checked ▲

b First names of your dependent children who lived with you _____   Enter number ▲

c Number of other dependents (from line 27) . ▲

d Total (add lines 6a, b, and c) . . . . . ▲   Enter number of boxes checked ▲

e Age 65 or over . . ☐ Yourself ☐ Spouse   Blind . . . ☐ Yourself ☐ Spouse

7 Total (add lines 6d and e) . . . . . . ▲

8 **Presidential Election** ▲ Do you wish to designate $1 of your taxes for this fund? . . ☐ Yes ▨ ☐ No   Note: If you check the "Yes" box(es) it will not increase your ___ your return.
If joint return ___ ich to designate $1? . .

FIGURE A-1 Portion of the 1975 federal individual income tax form containing residence question.

In an effort to assist the Census Bureau in assigning geographic codes, the Internal Revenue Service asked a residence question on the 1972 and 1975 tax returns. Complete responses to the questions were received for over 70 percent of the returns in 1972 and for over 95 percent of the returns in 1975.[14] The relevant portion of the 1975 income tax form is reproduced as Figure A-1. While the information from the residence questions allowed assignment of geographic codes to the tax returns for 1972 and 1975, assignment of geographic codes is necessary for every year. Geographic codes also need to be assigned to those returns with incomplete responses to the residence question. In order to work with limited information, the Census Bureau adopted the following imputation procedures (see Galdi (1978) for more details).

For each year for which the residence question was asked, a geographic "coding guide" was created. These guides relate the responses to the residence questions with the mailing addresses. In particular, each residence response is assigned a geographic code. Each mailing address is also coded to an address "key" identifying the state, zip code, first seven letters of post office name, and address type (numeric, rural, post office box, or other). For each key, the distribution of geographic codes corresponding to residence responses is observed. For example, suppose that for a given key, the residence responses on 1975 tax returns containing mailing addresses corresponding to that key were distributed as 84.12 percent inside the limits of city X in county Y, and 15.88 percent in county Y but outside the city limits. For each key, the observed distribution is used to assign "probability codes" to mailing addresses corresponding to that key. In other words, the probability codes are geographic codes that are randomly assigned to address keys, where the probability that any particular geographic code is assigned to a given key equals the observed proportion of geographic codes appearing on tax returns with that key. For the key in the above example the geographic code for city X in county Y would be assigned the probability .8412, and the geographic code for the "balance of county" and for county Y would be assigned the probability .1588.

Probability codes are used as surrogate geographic codes when the latter are not available. For tax returns in years other than 1972 or 1975 the probability codes are assigned according to the observed distribution for the most recent year for which the coding guide is available (currently 1975).

---

[14]The 1972 tax forms contained the residence question on the second page, while the questions for 1975 appeared at the top of the first page. The questions were also worded differently.

Probability codes are used in classifying matched pairs of tax returns as inmigrants, outmigrants, or nonmovers for a subcounty unit. For estimating migration over 1976 to 1978 (using 1975 and 1977 tax returns) the procedure was as follows:

1. The mailing addresses on the pair of matched returns are compared. If the address keys are the same or if other parts of the mailing addresses match, the persons represented by exemptions on the returns are classified as nonmovers.

2. If the mailing addresses do not match, the persons represented by exemptions on the returns are classified as inmigrants to the subcounty unit by using the geographic code (or probability code) for the later (1977) tax return and as outmigrants from the subcounty unit using the geographic code (or probability code[15]) for the earlier year (1975).

4.1d(2) *Use of Tolerance Levels* Another difference between the calculation of IRSRAT at the county level and at the subcounty level is the use of tolerance intervals to stabilize the values of IRSRAT for certain subcounty units. If a place with fewer than 20,000 people had a coverage ratio (ratio of exemptions on matched individual income tax returns to nongroup quarters base year population) falling outside a tolerance interval of 66 percent to 150 percent of the county coverage rate, IRSRAT for the place was set equal to IRSRAT for one of two larger areas. For estimates prior to 1977, if IRSRAT for the place was within 10 percent of the county IRSRAT, it was equated to the county IRSRAT.[16] Otherwise, IRSRAT for the place was set equal to IRSRAT for the ensemble of all places under 20,000 population in the county whose coverage rates fell inside the tolerance interval. The procedure now practiced uses the latter "ensemble" rate, unless it differs from the county rate by more than 10 percent (of the county rate), in which case the county rate is used. These stabilizations are invoked because, in the case of smaller areas, unusual coverage rates are often a symptom of geographic coding problems arising from post office consolidations, new incorporations or annexations, places split between counties, and distinct places possessing identical names (see Bureau of the Census (1980) or Healy (1978) for further discussion).

Net migration is estimated as

$$\text{NETMIG}(s, t; i, j, k) = \text{IRSRAT}(s, t; i, j, k) \times \text{MIGBASE}(s, t; i, j, k),$$

---

[15] Probability codes were used for the 5 percent of the 1975 returns for which complete responses to the residence question were not available. Probability codes were also used for 4.4 percent of the 1975 returns that were believed to contain reporting or coding errors.

[16] However, if the difference was within 5 percent of the county IRSRAT or if the difference in net migrants was less than 10, the original IRSRAT for the subcounty unit was not replaced.

where

$$\text{MIGBASE}(s, t; i, j, k) = \text{NGQPOP}(s; i, j, k)$$
$$+ \tfrac{1}{2}\{\text{BIR}(s, t; i, j, k) - \text{DEA}(s, t; i, j, k) + \text{IMM}(s, t; i, j, k)\}$$

and NGQPOP($s; i, j, k$) is the non-group quarters population at date s in subcounty unit k of county j, state i.

The discussion of error in using AR (see section 2.9 above) is relevant here for the subcounty estimates as well as for the state estimates.

4.1e *Immigration From Abroad Over (s, t]:* IMM($s, t; i, j, k$)

For every place whose 1970 population was at least 100,000, data on the number of immigrants from abroad are provided by the Immigration and Naturalization Service. Immigrants from abroad for the balance of the state (i.e., the state excluding places of 100,000 or more) are apportioned among places of less than 100,000 according to the number of persons of foreign birth counted there in the 1970 census.

4.1f *Military Barracks and Other Group Quarters Population:* MILBAR($s, t; i, j, k$) *and* IC($s, t; i, j, k$)

Information on special populations is gathered on an annual basis by members of the Federal-State Cooperative Program (FSCP). The Census Bureau has requested that the FSCP members obtain, at a *minimum*, data on (1) military barracks with over 100 people and (2) any other special population comprising at least 500 persons and at least 2 percent of the area's population.

The extent of data collected varies widely from state to state. Some FSCP members keep track of just points 1 and 2, while others obtain data on even the smallest group quarters populations. The group quarters populations considered are members of the armed forces living in military barracks, inmates of prisons, inmates of long-term hospitals, and, as a proxy for college students living in dormitories, college students enrolled in full-time programs. (Further details may be found in Bureau of the Census (1980).)

4.1g *Annexations and New Incorporations*

In January of each year the Census Bureau conducts the Boundary and Annexation Survey to determine whether there have been any boundary changes or governmental reorganizations (incorporations or disincorporations) during the preceding calendar year. The units of government surveyed include county governments and the governments of incorporated places. From 1971 to 1977 the Census Bureau did not include incorporated places with population under 2,500 in this survey. Beginning January 1978, however, all incorporated places were surveyed as well. The reason for the increase in the frame is related to the conduct of the 1980 census. Information about unincorporated places (townships) is provided by the underlying counties in which the places (townships) are located.

The procedures for adjusting the population estimates to reflect boundary changes will be described, first for areas of at least 5,000 population[17] and then for the remaining areas. Areas of at least 5,000 population that have undergone boundary changes are identified by the Boundary and Annexation Survey. For these areas the Census Bureau performs what is called a "separation": the 1970 population of the annexed or de-annexed area is computed from the 1970 census records. Prior to 1977 the Bureau's rule was that the postcensal population estimates would be recomputed to reflect boundary changes only if the 1970 population of the annexed (or de-annexed) area exceeded 5 percent of the 1970 population of the annexing area. At present, however, postcensal population estimates for all areas of at least 5,000 population are recomputed to reflect any new separations, such that[18] (1) the boundary changes involved new geography, e.g., a place in one township or county was annexed into another township (or county), (2) the 1970 population of the annexed (or de-annexed) area was at least 100,[19] or (3) a boundary change in a previous year had resulted in change of at least 5 percent in the area's population estimate. Currently, a separation is performed for an area of at least 5,000, provided that the area's estimate[20] of the population of the annexed (or de-annexed) area, as reported in the Boundary and Annexation Survey, is at least 5 percent of the 1970 population of the annexing area.

To recompute the population estimate for an area undergoing boundary changes, the Census Bureau attributes to the annexed (or de-annexed) area the estimated growth rate for the annexing area and then adds (or subtracts) the annexed (or de-annexed) area's population estimate to the annexing area's estimate.

Prior to 1977, population estimates for areas whose population numbered under 5,000 were not recomputed to reflect boundary changes (ex-

[17] These areas include both those with at least 5,000 population counted in the 1970 census and those with postcensal population estimates of at least 5,000.

[18] The rules are not rigid, and the postcensal population estimates are recomputed in other cases as well.

[19] In practice, this rule is not strictly applied, and many smaller separations are also taken into account.

[20] This estimate usually refers to current population.

cept in unusual cases). The rule is now that estimates are recomputed to reflect boundary changes for an area of population under 5,000 if the area requests and agrees to pay for the separation. The procedures for recomputing the estimates for areas under 5,000 are the same as those described above for use in areas of at least 5,000 population.

Regardless of the size of the area, updates by the AR method for the area in later years do not have to be modified to account for the boundary changes, because the additional population is reflected in the estimate of MIGBASE used to multiply the migration rate.

4.2 ADJUSTMENT OF ESTIMATES AND USE OF SPECIAL CENSUSES

The AR estimates of subcounty populations are scaled to sum to the county totals. The procedure is analogous to that described for county estimates in section 3.8 above. In a few instances, estimates of subcounty populations prepared by the state are also used by the Census Bureau. In such cases these estimates are scaled to sum to the county totals and then averaged with the AR estimates.

One final adjustment procedure remains. When a recent special census tally of subcounty population is available, it replaces the AR estimate of population or average of AR and state-prepared estimate, hereafter called "AR estimate." To force the total of the subcounty (county) estimates to sum to the county (state) totals, the "rake/float" procedure is used:

1. If the sum of 1970 populations of places in a county receiving a special census is at least one third of the 1970 county population, the sum of the differences between the AR estimates and the special census estimates is added ("floated") to the county total.

2. If the sum of 1970 populations for places in a county receiving a special census is less than one third of the county total, but the sum of the differences between the AR estimates and the estimates from special censuses exceeds in absolute value 3 percent of the county total, the excess over the 3 percent is added ("floated") to the county total, and the remainder (= 3 percent of the county total) is distributed in proportion to estimated population ("raked") over the areas in the county that did not have a special census.

3. If neither point 1 nor point 2 applies, then the sum of the differences between the AR estimates and the estimates from special censuses is distributed proportionately ("raked") over areas within the county that did not have special censuses.

The "county total" referred to above is the preliminary, or ORS, county

estimate described in section 3.1 above. For counties the rake/float procedure is analogous to that just described for places. Changes in the subcounty or county estimates brought about by raking, floating, and scaling
to county or state totals are all attributed to the net migration components.

4.3 SOURCES AND STRUCTURE OF ERRORS

Geographic coding is a major source of error in estimating subcounty
migration by the administrative records method. As has been noted, problems arise because the mailing address on a tax return is often insufficient
for determining in which unit of local government the filer resides. In
many cases the residence of the filer is not the same as the mailing address. For example, in many areas, people living outside the town limits
receive mail at post office boxes within the town limits. In a significant
number of cases the Census Bureau is unable to assign a mailing address
to a unique subcounty unit because zip codes, street addresses, and post
office jurisdictions often span geopolitical units. Also, some places lie in
several counties, and the mailing address will not indicate to which county
the address belongs, nor will the mailing address indicate whether it lies
inside or outside the city limits.

Using information obtained from special questions on residence appearing on the tax forms for 1972 and 1975, the Census Bureau constructed coding guides, which were used to assign tax returns to places of
residence on the basis of reported mailing addresses. Errors arise from use
of this coding guide as well. First, there are response errors to the questions on residence. The response rate to the question in 1975 was 95 percent, but there were also errors in the responses obtained. Healy (1978)
discusses errors in the responses to the question, such as a tendency for
some people living outside town limits to report their residence as being
inside the limits. Other response errors occur in connection with new incorporations, annexations, boundary changes, places straddling different
geographic units, or places in different counties possessing identical
names.

A second, more serious source of error is the use of the coding guide to
assign geographic residence codes to tax returns for other years than those
in which the question on residence is asked. If such a year is close to the
year when the coding guide was created (i.e., a year for which the question
on residence was asked), the chances of error are probably minimal. However, as the length of time between the year to be coded and the year the
coding guide was created increases, the coding guide will become more
and more seriously outdated because of boundary changes, changes in

mailing addresses caused by postal reorganization, and population growth.

The administrative records method rests on the assumption that the matching of tax returns for two separate years on the basis of social security number can yield migration rates that are representative of the whole population. The data underlying these computed rates obviously do not apply to (1) persons who do not file a tax return (or are not claimed as an exemption) at all or (2) persons (or dependents of persons) who filed a tax return in only one of the two years. There is some question whether the migration patterns of these people are similar to those of the population covered by the tax returns (Lowe et al., 1974; Mann, 1978). In addition, many persons claimed as exemptions—college students and in some cases children of divorced parents—do not reside with the person claiming them as an exemption.

For areas with population over 5,000, population changes caused by boundary changes are not as a rule reflected in the postcensal estimates when these changes are estimated by the annexing area to be less than 4½ percent of the area's 1970 population. For a large area this annexed area may contain a large number of people, but if the estimated ratio of the annexed area's population to the annexing area's population is under 4½ percent, no separation will be performed. Population changes resulting from boundary changes to areas with population under 5,000 are not reflected in the estimates unless the area requests and pays for a separation. For those areas undergoing boundary changes but not receiving separations, population changes arising from boundary changes will be detected only through the matches of tax returns. In the matching process, however, a person, not a recent migrant into the annexed area, will be treated as a nonmover and hence not reflected in the estimate of population change. For a resident of the annexed area who is a recent migrant, determining geopolitical unit of residence presents severe problems.

Estimation of births and deaths for places of population under 10,000 is unavoidably problematic because the tabulations of births and deaths for many of these areas are not available.

In summary, estimation of all components of population change is more difficult at subcounty than at county or higher levels. The overall extent of the errors in AR subcounty estimates is discussed in Part 2 of this report. Little is known, however, about the relative sizes of the errors in the estimates of the various components.

SPECIAL NOTE: GUIDE TO NOTATION AND CONVENTIONS

The following list indicates the locations of the definitions of various notation and conventions used in this appendix:

Notation		*Section*
ADJB	adjusted estimate of births	4.1b
AR	administrative records method	2.1
B	estimated births	4.1b
BIR	births	2.2a
CM II	component method II	2.1
DLAC	driver's license address change method	3.2
DEA	deaths	4.1
DEAEW	deaths to "elderly" whites	4.1c
DEAEWRAT	death rate for "elderly" whites	4.1c
DEAY	deaths to "young"	2.2a
DEAYW	deaths to "young" whites	4.1c
DEAYWRAT	death rate for "young" whites	4.1c
d_{ra}	national death rate for race r, age a	3.4c
ENROL	number of children enrolled in grades 1–8	2.2e
EXSCLPOP	expected school-age population	2.2e
FEMIGYRAT	migration rate for "young" females	2.2e
FR	fertility rate for females aged 15–39	4.1b
F1539	females aged 15–39	4.1b
GQPOPY	net change of group-quarters "young"	2.2a
HUM	housing unit method	3.3
IC	special populations other than military barracks residents, i.e., institutional and college	4.1
IMM	immigrants from foreign countries	3.7, 4.1
INS	number of tax exemptions classified as immigrants	3.7, 4.1d(1)
IRSRAT	migration rate calculated from IRS tax returns	3.7, 4.1d(2)
MEDCARE	number of Medicare enrollees	2.2b
MIGBASE	population base for multiplying a migration rate	3.7, 4.1d(2)
MIGERAT	migration rate for "elderly"	4.1c
MIGYRAT	migration rate for "young"	4.1b
MILBAR	population living in military barracks	4.1
NBY	non-barracks "young"	4.1c
NETMIG	net immigration of non-group quarters residents	4.1
NETMOVY	net movement of young from military group quarters to non-group quarters	2.2a, 3.7
NGQMIGY	net migration of non-group quarters "young"	2.2a
NGQMIGYRAT	migration rate for the non-group quarters "young"	2.2e
NGQPOP	non-group quarters population	3.7
NGQPOPY	non-group quarters "young"	2.5d
NONMOV	number of tax exemptions classified as nonmovers	3.7
OUTS	number of tax exemptions classified as outmigrants	3.7
POPE	"elderly" population	2.2a
POPY	"young" population	2.2a

Notation   *Section*

PU1	proportion of population under 1 year of age	4.1b
RC	ratio-correlation method	2.1
RESPOP	resident population	2.2a
SCLBIR	school-age children born since the last census	2.2e
SCLCHT	cohort of school-age children	2.2e
SCLDEA	deaths to school-age children	2.2e
SCLMIGRAT	migration rate of school-age children	2.2e
SCLPOP	school-age population	2.2e
U1	population under 1 year of age	4.1b
$(T_1, T_2]$	time period since T, up to and including T_2	2.2a

Conventions

elderly	population aged 65 or over on estimate date	2.1
ORS	population estimate used by Office of Revenue Sharing	3.1, 4.1a
preliminary	second set of county population estimates	3.1
provisional	first set of county population estimates	3.1
revised	third set of county population estimates	3.1
young	population aged less than 65 on estimate date	2.1

REFERENCES

Bureau of the Census (1973a) *Census of Population and Housing: 1970, Evaluation and Research Program PHC(E)-4.* Estimates of Coverage of Population by Sex, Race, and Age: Demographic Analysis. Washington, D.C.: U.S. Department of Commerce.

Bureau of the Census (1973b) *Census of Population and Housing: 1970, Evaluation and Research Program PHC(E)-7.* The Medicare Record Check: An Evaluation of the Coverage of Persons 65 Years of Age and Older in the 1970 Census. Washington, D.C.: U.S. Department of Commerce.

Bureau of the Census (1973c) *Census of Population and Housing: 1970, Evaluation and Research Program PHC(E)-2.* Test of Birth Registration Completeness 1964 to 1968. Washington, D.C.: U.S. Department of Commerce.

Bureau of the Census (1973d) *Federal-State Cooperative Program for Population Estimates, Federal-State Cooperative Program for Local Population Estimates: Test Results—April 1, 1970.* Current Population Reports, Series P-26, No. 21. Washington, D.C.: U.S. Department of Commerce.

Bureau of the Census (1974) *Estimates of the Population of States and Components of Change, 1970 to 1973.* Current Population Reports, Series P-25, No. 520. Washington, D.C.: U.S. Department of Commerce.

Bureau of the Census (1976) *Population Estimates and Projections, Estimates of the Population of States with Components of Change, 1970 to 1975.* Current Population Reports, Series P-25, No. 640. Washington, D.C.: U.S. Department of Commerce.

Bureau of the Census (1978a) *Special Studies, Population Estimates by Race, for States: July 1, 1973 and 1975.* Current Population Reports, Series P-23, No. 67. Washington, D.C.: U.S. Department of Commerce.

Bureau of the Census (1978b) *Population Estimates and Projections, Annual Estimates of the Population of States, July 1, 1970 to 1977.* Current Population Reports, Series P-25, No. 727. Washington, D.C.: U.S. Department of Commerce.

Bureau of the Census (1980) *Population and Per Capita Money Income Estimates for Local Areas: Detailed Methodology and Evaluation.* Current Population Reports, Series P-25, No. 699. Washington, D.C.: U.S. Department of Commerce.

Bureau of Labor Statistics (1978) *Employment and Earnings* 25(5)(May).

Cavanaugh, F. J. (1977) Memorandum to J. Glynn: Computation of July 1, 1976 Population Estimates by the Administrative Records Method. July 26, 1977. Bureau of the Census, U.S. Department of Commerce.

Galdi, D. (1978) Memorandum for the Record: IRS Geo-Coding Methodology. November 29, 1978. Bureau of the Census, U.S. Department of Commerce.

Healy, M. K. (1978) Administrative Records—Subcounty Population Estimates. Paper presented at the Census Tract Conference, U.S. Department of Health, Education, and Welfare, Annapolis, Md., Nov. 15–17, 1978.

Irwin, R. (1978) Aggregate Medicare Enrollment by Age, Sex, and Race as a Resource in Analyzing Demographic Change for Local Areas. Paper presented at the National Bureau of Economic Research Workshop on Policy Analysis with Social Security Research Files, Williamsburg, Va., March 15–17, 1978.

Lowe, T., Walker, J., and Weisser, L. (1974) Evaluation of Estimates Based on Income Tax Returns. Paper presented at the Bureau of the Census Meeting, Reno, Nev., Dec. 16, 1974.

Mann, E. (1978) Problems with data in selected formula funded programs as applied to New York City. Pp. 114–119 in *1978 Proceedings of the Section on Survey Research*

Methods of the American Statistical Association. Washington, D.C.: American Statistical Association.

Morrison, P. (1971) *Demographic Information for Cities: A Manual for Estimating and Projecting Local Population Characteristics.* RAND Report R-618-HUD.

Pittenger, D., Lowe, T., and Walker, J. (1977) Making the Housing Unit Population Estimation Method Work: A Progress Report. Paper presented at the annual meeting of the Population Association of America, April 22, 1977, St. Louis, Mo.

Purcell, N. J., and Kish, L. (1979) Estimation for small domains. *Biometrics* 35:365–384.

Rasmussen, N. (1975) The use of drivers license address change records for estimating interstate and intercounty migration. Pp. 16–22 in Bureau of the Census, *Intercensal Estimates for Small Areas and Public Data Files for Research.* Small-Area Statistics Papers Series GE-41 No. 1. Washington, D.C.: U.S. Department of Commerce.

Starsinic, D., and Zitter, M. (1968) Accuracy of the housing unit method in preparing population estimates for cities. *Demography* 5(1):475–484.

van der Vate, B. (1978) Memorandum for Federal State Cooperative Program Participants. Subject: Calculation of Rates and Factors in the Component Method II. September 28, 1978.

Warren, R., and Peck, J. (1975) Emigration from the United States: 1960 to 1970. Paper presented at the annual meeting of Population Association of America, April 17–19, 1975, Seattle, Wash.

APPENDIX
B

Postcensal Per Capita Income Estimation Methods of the Census Bureau: Summary

DONALD E. PURSELL and
BRUCE D. SPENCER

The Census Bureau defines the per capita income of an area as the mean or average total money income of residents during the preceding year. Thus the 1974 per capita income of an area is the mean income of the population on April 1, 1974, during the calendar year 1973. Total money income is the sum of six components: wage and salary income; non-farm proprietors' income; farm proprietors' income; social security and other retirement income payments; public transfer payments (assistance payments); and other income, including interest dividends, unemployment insurance, etc.

To estimate postcensal per capita income for states and counties, the Census Bureau makes separate postcensal estimates of each of the six components of total money income, adds them, and then divides the sum by the estimate of postcensal population.[1] Postcensal estimates of sub-county per capita income are obtained by direct estimation of the rate of change in per capita income and the application of this rate to the 1970 census estimate of per capita income. These methods are described more fully below; see also Bureau of the Census (1980).

STATE UPDATES

To estimate postcensal per capita income for states, the Census Bureau updates the estimate of total money income and then divides by the post-

---

[1] As discussed below, county wage and salary income is updated on a per capita basis.

censal population estimate (see Appendix A). In updating the money income estimate, updates for each of the six components are made separately and then summed. In this section we consider postcensal estimates
for 1975 (that is, 1974 per capita income).

For wage and salary income the Census Bureau uses data from the Internal Revenue Service (IRS). The ratio of wage and salary income for 1974
to that for 1969 is estimated by the ratio of wage and salary income
reported to IRS for 1974 to that reported for 1969. This ratio is then multiplied by the estimate of 1969 wage and salary income from the 1970 census, yielding the postcensal estimate of wage and salary income.

Updates for the other five types of income are obtained from the Bureau
of Economic Analysis' personal income estimates. The procedure is similar to that for the wage and salary updates, but an extra adjustment is
used because the BEA personal income figures are based on the midyear
(July 1) population for the respective year. Thus personal income for 1974
refers to income of the 1974 population in 1974, while the Census Bureau's
total money income for 1974 refers to income the 1975 population received
during 1974.

The ratio of the public assistance component of total money income for
1974 to that for 1969 is estimated by

$$\frac{\dfrac{1974 \text{ BEA public assistance income}}{1974 \text{ population}} \times 1975 \text{ population}}{\dfrac{1969 \text{ BEA public assistance income}}{1969 \text{ population}} \times 1970 \text{ population}} \; .$$

To obtain the postcensal estimate of public assistance income, this ratio is
multiplied by the estimate of 1969 public assistance income provided by
the 1970 census. The other components of income—net nonfarm self-
employment, net farm self-employment, social security and railroad
retirement, and other income—are estimated analogously to public assistance income.

Total money income for 1974 is estimated as the sum of the six income
components. Each state's total money income is divided by the postcensal
estimate of the April 1, 1975, population of the state. This population estimate is calculated as one-fourth of the July 1, 1974, postcensal population
estimate plus three-fourths of the July 1, 1975, estimate.

COUNTY UPDATES

For four of the six components of income, postcensal estimates of total
money income at the county level are obtained as at the state level. Wage

and salary income and farm self-employment income are handled differently.

The wage and salary updates are done on a per capita basis in order to minimize the effect of possible errors in the geographic coding of tax returns. The ratio of per capita wage and salary income for 1974 to that for 1969 is estimated by the ratio of the average reported wage and salary income per exemption on the 1974 IRS tax forms to the average reported wage and salary income per exemption on the 1969 IRS tax forms. Postcensal wage and salary income is then estimated as the product of that ratio, the estimate of 1969 per capita wage and salary and income provided by the 1970 census, and the estimate of the April 1, 1976, population.

There are two major problems in obtaining estimates of farm income. First, county farm income is notoriously volatile, capable of major, sharp, year to year changes. These changes may be either understated or overstated by the data used to measure them. Second, the problems of comparability between BEA and Census Bureau estimates for farm self-employment income are severe. In particular, BEA estimates tend to show considerably more annual variation than do estimates from censuses and surveys. For these reasons the Census Bureau initially prepares two estimates of postcensal farm self-employment income, a "net" farm income estimate and a "gross change" farm income estimate, and then uses those estimates to derive a "constrained net estimate" of farm self-employment income. The "net" farm income estimate is derived as the sum of (1) the 1970 census estimate of 1969 farm self-employment income and (2) the dollar change in BEA farm self-employment income plus land rent. The "gross change" farm income estimate is obtained by applying the ratio of 1974 BEA farm receipts to 1969 BEA farm receipts (adjusted to account for the July 1 reference base of BEA estimates) to the 1970 census estimate of 1969 farm self-employment income and adding to this the sum of (1) the dollar change in BEA land rent and (2) the 1970 census estimate of 1969 land rent.[2] The constrained net estimate is then calculated as the median of three quantities: net farm self-employment income, 80 percent of the gross change estimate of farm self-employment income, and 120 percent of the gross change estimate of farm self-employment income. This constrained net estimate is used as the postcensal estimate of farm self-employment income. Approximately 25–30 percent of the county estimates are directly affected by the constraints, that is, are based on the gross change rather than the net estimate.

The postcensal estimates of the six income components are then added

[2] The 1970 census estimate of 1969 land rent is estimated as the farm self-employment income of nonfarmers.

to yield an estimate of total money income. After adjustment to the state total money income, the county estimate is divided by the appropriate estimate of population, yielding the postcensal estimate of county per capita income. This procedure was followed for 1972 (initial and revised) and for 1974 (initial) per capita income estimates; however, additional constraints were incorporated into the procedure beginning with the 1974 (revised) and the 1975 (initial) per capita income estimates.

In the new procedure, total money income is decomposed into two parts, adjusted gross income (AGI) and transfer income (TI). The latter, TI, is composed of social security income, public assistance income, and part of "other" income, such as unemployment and veterans' payments; the former, AGI, is the rest of total money income. Estimates of AGI are determined by adding the component estimates derived above, using BEA estimates to allocate "other" income between AGI and TI. The ratio (for 1974 income)

$$A = \frac{1974 \text{ county per capita AGI}/1969 \text{ county per capita AGI}}{1974 \text{ state per capita AGI}/1969 \text{ state per capita AGI}}$$

was computed, where per capita AGI for year 1974 (1969) is the estimated AGI for 1974 (1969) divided by the estimate of population for 1975 (1970). A similar ratio was computed from income reported on tax forms,

$$B = \frac{1974 \text{ county IRS AGI per exemption}/1969 \text{ county IRS AGI per exemption}}{1974 \text{ state IRS AGI per exemption}/1969 \text{ state IRS AGI per exemption}},$$

where IRS AGI per exemption refers to the ratio of the total AGI income reported on IRS individual income tax forms to the number of exemptions claimed on the tax forms. The constrained estimate of county AGI was then obtained as the median of A, $B + 0.25$, and $B - 0.25$. Total money income is then recomputed by adding the constrained estimate of AGI to the estimate of TI. The estimates of county total money income are scaled to sum to the state estimate of total money income. Per capita estimates are calculated by dividing the totals by the respective population estimates.

SUBCOUNTY UPDATES

The derivation of subcounty estimates roughly parallels that of the county estimates. Significant differences do exist, however:

1. The per capita estimates are updated directly rather than by separate updating of total money income and population.

2. The BEA estimates are not available for subcounty units—for income components not measurable from IRS records, county per capita estimates are applied to all subcounty units.

3. Many constraints are employed to damp changes in the estimates.

4. The 1970 census estimates were modified to reduce sampling variability and to account for boundary changes and annexations.

5. Changes in subcounty per capita income are estimated in multiple increments rather than single increments. Thus the change from 1969 to 1974 is estimated by the sum of the changes from 1969 to 1972 and 1972 to 1974; the procedure is similar to the administrative records method used in population updates (see Appendix A).

6. There is a complicated adjustment of subcounty estimates to county totals.

For 1972 per capita income updates, the observed rate of change in IRS AGI per exemption (as defined earlier) and the observed rate of change in BEA per capita county transfer income were applied to 1970 census estimates of these components for 1969. The procedures for 1974 (1975) updates were similar except that the rates of change referred to the period 1972–1974 (1974–1975) and the base estimates were for 1972 (1974) rather than for 1969. For simplicity, only the 1972 updating procedure is described here; the other procedures are analogous. (For more details, see Herriot (1978) and Bureau of the Census (1980).)

The first stage of the updating procedure consists of four operations to the 1969 base per capita income figures.

1. The 1969 per capita income figures for areas experiencing annexations and boundary changes from 1969 to 1972 were modified to adjust for any resultant changes in per capita income.

2. To reduce the effect of sampling variance in the 1970 census estimates for places of population under 1,000, a weighted average of the 1970 census estimates and regression estimates was used. The weights were derived by applying James-Stein techniques; see Fay and Herriot (1979).

3. The 1969 income estimates were decomposed into transfer income and adjusted gross income so that the IRS data on adjusted gross income could be used.

4. After the above three procedures were carried out, the sum of the estimates for subunits of geography might have differed substantially from "independent" estimates of the total. An iterative adjustment proce-

dure was used to simultaneously control subcounty estimates to the county total and force the sum of the estimates for each of several size classes of places to add to the statewide totals for the size classes.

The second stage of the updating procedure consisted of estimating and applying the rates of change in per capita income.

1. The IRS data were adjusted for annexations and boundary changes.
2. To protect against severe errors in data, a host of edits and constraints (at least 11) were imposed. Those constraints had the effect of restricting estimates of rates of change for geographical subunits to be close to the corresponding countywide or statewide average rate of change.
3. After the edits and constraints were imposed, the rates of change in IRS AGI per exemption and in BEA per capita countywide transfer income were applied to the respective estimates for the base period.
4. After those updates of per capita adjusted gross income and transfer income were made, they were separately forced to sum to estimates of the total. The procedures are similar to those used for controlling the estimates for the base period.

After controlling to totals, the final per capita income estimate is obtained by dividing the sum of per capita adjusted gross income (AGI) and per capita transfer income (TI).

REFERENCES

Bureau of the Census (1980) *Population and Per Capita Money Income Estimates for Local Areas: Detailed Methodology and Evaluation.* Current Population Reports, Series P-25, No. 699. Washington, D.C.: U.S. Department of Commerce.

Fay, R., and Herriot, R. (1979) Estimates of income for small places: An application of James-Stein procedures to census data. *Journal of the American Statistical Association* 74(366):269–277.

Herriot, R. (1978) Updating per capita income for general revenue sharing. Pp. 8–15 in Bureau of the Census, *Interrelationships Among Estimates, Survey, and Forecasts Produced by Federal Agencies.* Small Area Statistics Papers, Series GE-41, No. 4. Washington, D.C.: U.S. Department of Commerce.

The Role of Judgment in Postcensal Estimation

C

BRUCE D. SPENCER

Judgment underlies every application of statistical theory and methodology. A statistical procedure may be justified by well-defined assumptions, but the applicability of those assumptions in any given situation is determined and decided by judgment. In demographic estimation, judgment is especially pervasive. The purpose of this paper is to examine how judgment enters into the formulation and use of the postcensal estimation methodology.

One can distinguish between a stated protocol for assembling and analyzing classes of individual units of information and a less rigidly stated, more flexible approach. The former may be termed less subjective, i.e., less dependent on judgment. There are degrees of subjectivity. For example, one procedure might treat all units of information in precisely the same manner but with provision for exceptional treatment of units possessing specified unusual characteristics. Another procedure might rigidly specify the process for every stage of analysis of individual units except the final one, which may depend on judgment. Yet another procedure may involve interpretation of a body of data that rests almost entirely on the judgments of one or more experts. Perhaps one key to distinguishing fixed methods (objective techniques) from subjective ones (judgment) is the extent to which a given procedure is *reproducible* by a second person or team.

One can identify several ways in which judgment enters into making postcensal estimates: (1) in formulating methodology, (2) in deciding to edit (accept, reject, or modify) data, (3) in deciding when a method or estimate should be rejected in a particular instance, and (4) in modifying methods.

1. The role of judgment in formulating methodology is indispensible not only for demography but for all scientific endeavors. Bayesian decision theory is the most explicit in its use of subjective notions, but all statistical analysis has some degree of subjectivity, e.g., definition, choice of models, mechanics of estimation, selection of analytic techniques to guide inferences, presentation of evidence and conclusion.

2. Demographic methods draw upon data to produce estimates of parameters such as resident population or annual per capita income. Of course, data contain errors, and severe errors can cause serious inaccuracies in the estimates. Thus demographers at the Census Bureau screen the input data for possible large errors or outliers. If a piece of data coming in does not seem consistent with past trends or other current data, it is flagged. The decision to use a particular editing protocol or outlier-detection technique is usually a matter of judgment, although the use of a specified editing routine is most often governed by objective rules.

Objective rules are illustrated by the Census Bureau's use of tolerance intervals for estimation of births, deaths, and migration rates for sub-county units (see Appendix A, sections 4.1b–4.1d). If the data fail to meet explicit criteria, the data are rejected. For example, if too few persons in an area file tax returns, the migration information for the area is rejected. These rules are applied automatically, by computer.

The screening for outliers of school enrollment data used in component method II affords a good example of the use of judgment; CM II uses the calculated migration rate for school-age children in grades 1–8 as the primary basis for making inferences about the migration rate for the population as a whole (see Appendix A, section 2.2). For illustrative purposes, consider estimating the migration rate for these children 1 year after the last census. Suppose (for simplicity) that all school-age children are actually enrolled in school and suppose also that no deaths have occurred to these children during the year. Then the migration rate for the school-age population is estimated by

$$\frac{\text{ENROL}_1 - \text{EXPEC}_1}{\text{ENROL}_0},$$

where ENROL_1 is the number of school-age children reported enrolled in grades 1–8 for the current year; ENROL_0 is the number in grades 1–8 for the preceding (census) year; and EXPEC_1 is the number of children who would have been enrolled in grades 1–8 had there been no migration. The demographers at the Census Bureau calculate $ENROL_0$ and $EXPEC_1$ on the basis of census data but must use current data to estimate $ENROL_1$. To determine whether these current data are reliable, the demographers rely on their judgment. They will often consider several comparisons, such as comparing $ENROL_1 - ENROL_0$ with a historical time series of yearly differences in enrollment or comparing the difference between $EXPEC_1 - EXPEC_0$ and $ENROL_1 - ENROL_0$ with a historical time series of these differences. If the demographers perceive an abrupt change from a historical trend, they will flag the current enrollment data, $ENROL_1$, as an outlier. The data checks are not performed according to fixed rules, and thus they may not be replicable.

Although the use of judgment can be more complicated than the use of formal rules, some judgmental tests can be formalized. Since a formal test can be performed more quickly and more accurately, usually by computer, a greater variety of tests can be done. Precise specification of procedures (a prerequisite for objective tests) makes them amenable to statistical analysis, so that statistical properties, such as confidence limits, can be discovered. Knowledge of these properties could permit rankings of priority for data to be screened, so that if large amounts of data are suspect, the worst cases can be selected for early screening. Statistical analysis can also lead to the introduction of sophisticated improvements.

A recommended practice is to search for and use methods that tend toward objectivity when it can be approached, and to use subjective methods (judgment) only when the state of the art fails to provide satisfactory objective rules and techniques. The increased role of computers in data screening should supplement rather than replace the use of judgment. Freed from the necessity of making routine calculations, analysts can spend more time interpreting the results of the calculations and devising new kinds of tests.

Three things can happen to a piece of data flagged as an outlier: (1) an attempt is made to verify the data, (2) the piece of data is rejected and replaced by a substitute value, or (3) the datum is "trimmed." If the datum is verified, it is accepted; if verification attempts show the datum to be invalid, practice 2 is used. An example of practice 2 is the estimation of migration rates for subcounty units with fewer than 20,000 inhabitants (see Appendix A, section 4.1d). If the data for such a place do not satisfy certain formal criteria, the migration rate is set equal to either the county migration rate or the migration rate for all places in the county whose data did satisfy the criteria. The estimation of postcensal per capita income provides many illustrations of "trimming" (see Appendix B). For example, county farm self-employment income is estimated by "net farm self-employment income," provided the latter falls between the tolerance limits of 80–120 percent of "gross change farm self-employment income." If the net farm self-employment income falls outside these limits, it is trimmed to the nearest limit and then used as the estimate of farm self-employment income.

Current practice at the Census Bureau treats data flagged by computer by practice 2 or 3 above. Data flagged by judgment are treated by practice 1. This method is practical but not ideal. The number of separate data that are flagged by computer is so large that verification is not feasible for each. But the widespread use of practice 2 or 3 forces the data to reflect an often unreal stability.[1] Moreover, the tests now used to screen data for outliers search for data deviating from past trends. If in fact the underlying parameter changes but the data do not reflect this change, the data will not be flagged.

There is a need for more verification of data flagged by computer, instead of mere editing of the data. To verify all the flagged data would be prohibitively expensive; what is needed is a way to identify the data whose verification should receive priority. One useful approach would be to design and implement objective criteria to identify such data and assign them priority rankings. The decisions regarding which data should be verified could either be made objectively, solely on the basis of the assigned rankings, or subjectively, partly according to the rankings and partly according to other considerations.

3. Judgment is used to decide when a method or estimate should be rejected in a particular instance. A good example is the decision of when to stop incorporating information from a past special census into the estimates of population provided by CM II or the RC method (see section 3.11 of Appendix A).

4. The Census Bureau continually revises its methodology in minor and not-so-minor ways. The decisions to make the revisions are made on the basis of statistical evaluations, demographic logic (see Bureau of the Census, 1974), or professional judgment. These approaches are described and compared below.

In a statistical evaluation the estimates provided by a given method are compared to "benchmarks" (typically, estimates of high accuracy) such as decennial or special census counts. Quantitative measures of accuracy can then be computed to serve as a basis for evaluating the merits of the method. The benefits of this approach are objectivity and quantification. Properly performed, a statistical evaluation is not affected by the beliefs of

---

[1] Practice 2 typically assigns a large-area (county) rate of change to a component area (subcounty), and practice 3 shrinks the rate of change toward zero.

those performing the evaluation. Furthermore, statistical measures such as bias and standard deviation can be estimated. The major drawback in using statistical evaluations is the relative lack of representative benchmarks. Decennial census counts are of course available only every 10 years; a method that tested well for the 1960s may be poor for the 1970s. Special censuses are carried out only for a small proportion of areas, and the areas receiving them constitute a nonrandom sample (see section 2.1b of the report). An alternative consists of using low-precision benchmarks, which may be more readily available than high-precision benchmarks (see Appendix H). The regression-sample method is another effective alternative for counties (see section 3.2 of the report and Ericksen (1974)).

Use of logic involves replacing the assumptions underlying a method by more plausible assumptions. Whereas demographic logic considers the internal consistency and reasonableness of the methods, professional judgment focuses on the plausibility of the output of the methods. Such exercise of professional judgment is similar to statistical evaluations except that benchmarks are replaced by subjective estimates. The latter, of course, need not be based on introspection but can draw upon observations and current data not used by the method under consideration. For example, the Census Bureau's decision to drop births as a predictor variable in the ratio-correlation estimates of state populations in the 1970s was based on judgment (Bureau of the Census, 1974, p. 11):

For the 1960–1970 period and the 1950–1960 period as well, births had been one of the strongest indicators of population. ... However, some States [in the early 1970's] were in the process of removing restrictions on abortions in advance of the 1973 Supreme Court ruling. In these States, the decline in the number of births between 1970 and 1972 was much sharper than for the remainder of the Nation. As a result, the ratio-correlation estimate gave unrealistically low population estimates for these States. This was most apparent in the two largest States, California and New York.

The Census Bureau's decision not to use the composite method for estimating county populations in the 1970s was based on similar reasoning. Note that both the composite and ratio-correlation methods using births performed very well in the Census Bureau's tests of methods (see Bureau of the Census, 1973, 1974). If something is awry in the method or data, exercise of judgment may be the only recourse. The danger is that should the estimation method be accurately indicating unexpected trends, judgment may obscure perception of these. Where possible, statistical tests should be used to supplement or supplant this professional judgment. For example, the regression-sample method could have been used to justify the decisions just described. Use of error models (as discussed in

Appendix G) can improve the effectiveness of such subjective methods. Judgment is the means by which challenges to the Census Bureau's estimates are resolved (not necessarily presented). When decisions must be made for individual cases that may vary widely in circumstance, judgment may be the only reasonable method.

REFERENCES

Bureau of the Census (1973) *Federal-State Cooperative Program for Local Population Estimates: Test Results—April 1, 1979.* Current Population Reports, P-26, No. 21. Washington, D.C.: U.S. Department of Commerce.

Bureau of the Census (1974) *Estimates of the Population of States with Components of Change, 1970 to 1973.* Current Population Reports, P-25, No. 520. Washington, D.C.: U.S. Department of Commerce.

Ericksen, E. P. (1974) A regression method for estimating population change for local areas. *Journal of the American Statistical Association* 69(348):867–875.

Review Guide
for Population
Estimates

As was discussed in section 1.2c of the report, the Census Bureau sends its population estimates to the areas for their review before the estimates are published. The Census Bureau also sends a review guide, which is reproduced below.

REVIEW GUIDE FOR LOCAL POPULATION ESTIMATES

ESTIMATING METHOD

The population estimate shown in the enclosed notice was developed by use of a component procedure in which each of the components of population change (births, deaths, net migration, and special populations) were estimated separately. The estimates were derived in four stages, moving from the 1970 census as the base year to develop estimates for 1973, and in turn, moving from 1973 as the base year to derive estimates for 1975, from 1975 as the base year for 1976, and from 1976 as the base for 1977.

Natural change — Reported resident birth and death statistics were used, where available. These data were collected from State health departments and supplemented, where necessary, by data prepared and published by the U.S. Department of Health, Education, and Welfare, National Center for Health Statistics. For subcounty areas where reported birth and death statistics were not available from either source, estimates were developed by applying fertility and mortality rates.

Migration — Individual Federal income tax returns were used to measure migration by matching individual returns for successive periods. The places of residence on tax returns filed in the base year and in the estimate year were noted for matched returns to determine inmigrants, outmigrants, and nonmigrants for each area. A net migration rate was derived for each locality, based on the difference between the inmigration and outmigration of taxpayers and dependents, and was applied to a base population to yield an estimate of net migration for all persons in the area. Immigrants from abroad are added based upon data from the U.S. Immigration and Naturalization Service.[1]

Adjustment for special populations — In addition to the above components of population change, estimates of special populations were also taken into account. Special populations include persons who are residents of an institution, college, or military barracks. Data for these groups are collected from the specific institutions involved.

Other adjustments — In seven States (California, Florida, Oregon, New Jersey, Vermont, Washington, and Wisconsin) the subcounty estimates developed by this method were averaged together with estimates prepared by an agency in each State responsible for producing local population estimates. Special censuses were used in place of an estimate for localities where special censuses were taken close to the estimate date. The census results were adjusted to represent the population on July 1, 1977. Places which have had boundary changes since January 1, 1970 and before December 31, 1977, may have their 1970 census count and 1977 population estimate adjusted to reflect the population living in the annexed areas at the time of the 1970 census. There is a cost involved for the determination of the 1970 population in annexed areas, however, and an estimate of this cost can be provided by telephone. In places where this determination has already been made, the enclosed estimate reflects the adjustment.

COUNTIES

Estimates of the population of counties, independent cities, cities whose boundaries are coextensive with a county, and cities made up of more than one county, were developed by a technique that differs from that used for the subcounty places. The reason for this is the availability of more types of data sources at the county level enabling the derivation of estimates by more than one method.

The first technique is Component Method II. This procedure uses school enrollments to estimate the migration of persons under the age of 65. Births and deaths are tallied from reported county resident births and deaths to the population under age 65. The county population over age 65 is estimated based on the change in the number of Medicare enrollees. These two estimates by age are then added together to produce an estimate for the total county population.

---

[1] This number refers to legal immigrants only, since illegal immigrants cannot be enumerated or estimated accurately from this or other data sources.

BC-840
(2-6-79)

U.S. DEPARTMENT OF COMMERCE
BUREAU OF THE CENSUS

COUNTIES — Continued

The second technique is the Administrative Records method described earlier for subcounty places. The only variation at the county level is that the technique is specific for the population under age 65. The population over 65 is produced by the same method used in Component Method II. These separate estimates are combined to generate the total county figure.

The average population change between 1976 and 1977 for these two methods is added to the 1976 county estimate published in Current Population Reports, Series P-26. In approximately 15 States, additional data are available to permit the use of a third estimating technique that relies upon regression procedures to link shifts in local population with changes in related factors that are symptomatic of population change. A full discussion of all three methods can be found in the series P-26 reports and in series P-25, No. 640.

APPEALS AND CHALLENGE CONSIDERATIONS

The resulting figures for counties and local areas are an estimate of the population, not an actual count. A census of the entire U.S. population has not been conducted since 1970. Nonetheless, many public programs and planning activities require more up-to-date information.

Due to the nature of estimates, however, some error is always experienced in any technique used. The estimates produced by the Census Bureau for all levels of governments have undergone extensive testing and evaluation, with the results indicating acceptable error levels.[2] Locally prepared estimates also must be based upon thoroughly tested and recognized procedures. Even after thorough evaluation of the figures, challenges to the estimates should only be made when the differences between the Census Bureau estimate and locally derived figures are substantial enough to take into account expected estimation error levels.

Locally derived alternative estimates that are sent as a challenge should be accompanied by complete documentation describing in detail the derivation of the figure and the sources of the data used. For example, localities frequently rely upon the housing unit method for an alternative estimate. If a housing unit method estimate is sent, the following problems must be accounted for specifically in the documentation provided:

1. The building permits must be specific to your incorporated limits only.

2. Annual time series of residential permits and demolitions from 1970 to the 1977 estimate date must be supplied.

3. Estimates based upon units denoted by type (i.e., single-family and multi-family) are preferred, if data permit.

4. The data must relate to a July 1977, population estimate date only. Therefore, a time lag before the July 1977 date must be used to allow for the time between the issuance of permits and the completion of the units. A lag of 3 to 6 months is appropriate, depending upon local conditions.

5. Permits for commercial and home improvement projects should be removed from the data to reflect only residential units.

6. Demolitions and conversions to commercial uses should be registered and removed from the housing stock.

7. Vacancies must be accounted for.

8. Between 1970 and 1977, the U.S. average household size declined by 9 percent as a result of fewer births and an increase in one- and two-person households. Although the change will vary depending upon the size and type of community involved, any estimate based upon a housing unit method must take into account a household size change factor.

9. Nonhousehold populations that are living in group quarters should be accounted for separately. That is, they should be removed from the population totals in the initial year and replaced at the estimate date. These group quarters populations must consist of only those institutions having long-term housing facilities (e.g., college dormitory populations, inmates of Federal or State prisons).

---

[2] An evaluation of the methods is published in Current Population Reports, Series P-26, No. 21, and Series P-25, Numbers 740 to 789. A more detailed evaluation is forthcoming in Current Population Reports, Series P-25, No. 699.

APPEALS AND CHALLENGE CONSIDERATIONS — Continued

Although accuracy of the vacancy and population per household factors is critical in the housing unit method, the inventory of housing should not be overlooked as a potential source of error.

If utility data are used to estimate the number of occupied residential units instead of building permits, all of the above considerations must be accounted for, except for vacancies and the time lag needed for building permits. In addition, treatment of the following problems must be documented:

1. The coverage of the population by the utility must be evaluated against the 1970 household count, i.e., the number of housing units serviced by the utility in 1970 should be in general agreement with the number of occupied housing units enumerated in the 1970 census.

2. Master meters should be accounted for, and conversions from master meters to individual meters must be checked.

In order to obtain more accurate current information concerning the vacancy rates and population per household factors specific to local areas, some communities have conducted sample surveys. However, in such cases, it will be necessary to accompany the results with documentation specifying the sample design, the derivation of the sampling frame used, the assumed confidence limits and how they were developed, and an estimate of sample bias. Final computations should include a measure of the standard error. Other areas that are considering undertaking survey work as a part of their appeal should be in contact with us before initiating the project.

If any test of the estimating methodology has been made, the results should accompany the other challenge materials. This could consist of a comparison between the 1970 census count and an estimate of 1970, using the 1960 census as a starting point and your particular estimating method as the technique used to derive the 1970 estimate. This would enable us to better assure the accuracy of your particular technique.

Population projections are not suitable as challenge information since they do not reflect current data trends, but rather attempt to predict future change. Frequently projections are based on past growth patterns or on a series of assumptions concerning population change factors. Estimates, on the other hand, use current data series that are symptomatic of present population changes. Casual personal observations, "informed" opinions, and similar undocumented information cannot be used as a basis for an appeal.

Caution should also be taken that the population estimate conforms to the same definition of usual place of residence as is used in the decennial census. That is, temporary residents who live most of the year elsewhere should be excluded.

THE CHALLENGE PROCESS

Once a challenge is received by us, it goes through a detailed review. This process includes examination of the data series provided by the challenging locality together with a second detailed review of the data used in our estimating procedure. If it is impossible to resolve differences in the results based on the data series provided, any additional information available to us will be consulted and you may be contacted for further clarification and help.

If deficiencies are found in the information used by us in preparing the original population estimate, and if the data supplied by the challenging locality substantiate a different population figure, a change will be made. The revised estimate will be provided to the Office of Revenue Sharing (ORS) and other Federal agencies which use the population estimates in the distribution of Federal grant-in-aid funds, and you will be notified of the change. If a challenge is unsuccessful, it is often due to insufficient data for the challenge, the data supplied support the original population estimate, or the challenge materials are based on personal observation rather than firm support data and estimating techniques.

THE CHALLENGE PROCESS — Continued

If a challenge is unable to be resolved through the informal procedures described above, a State or unit of local government may request a formal hearing. Details for formal hearings are contained in regulations to be printed in the Federal Register during March 1979. The major provisions (1) stipulate that an informal challenge be filed no more than 180 days after release of the estimates, (2) require a locality to complete an informal review jointly with the Census Bureau before a formal hearing is allowed, (3) specify the appointment of a hearing officer to receive both written and oral evidence under oath, (4) allow for the cross-examination of both parties in the proceedings and of all witnesses, if requested, and (5) require that all action on challenges be completed within one year of the date of release.

In past years, a further and final resolution could be made by conducting a Federal special census. However, since field preparations are already underway for the 1980 national census, we are unable to contract for special censuses.

Please accept our thanks in advance for your cooperation and assistance in this review procedure. Also, please note that the population figure you will receive shortly from ORS may not reflect revisions made as a result of this challenge process. This is due merely to the timing of the ORS preliminary allocation work, and will be corrected before the final distributions through general revenue sharing and other Federal programs.

General Revenue Sharing Allocations and the Effects of Data Errors

BRUCE D. SPENCER

DETERMINATION OF GENERAL REVENUE SHARING ALLOCATIONS

General revenue sharing (GRS) allocations are determined according to data-based formulas. Application of the formulas is complicated and is performed by computer. The essentials of the procedure are described below; more complete discussions, in order of increasing detail, are found in the work of Nathan et al. (1975), U.S. Congress, Joint Committee on Internal Revenue Taxation (1973), Spencer (1980), and Bowditch et al. (1974). Descriptions of the various kinds of data input to the formulas are given by Office of Revenue Sharing (1973 et seq.).

The calculation of general revenue sharing allocations includes four major steps. First, allocations to the 51 state areas (the 50 states and the District of Columbia) are determined. Second, each state area's amount is split into two shares, a state government share and a statewide local government share. Third, each statewide local government share is partitioned among all county areas. Fourth, each local jurisdiction's allotment is calculated from the total available for the county area containing the jurisdiction. Local jurisdictions include county governments, township governments, Indian tribal councils, Alaskan native villages, and the governments of municipalities and places. The presence of maximum and minimum constraints causes the second, third, and fourth stages to be performed several times.

Two formulas are used to determine the allocations to state areas: a

"5-factor" and a "3-factor" formula. Because the "5-factor" and "3-factor" formulas originated with the House and Senate, respectively, they are also referred to as the House and Senate formulas. The allocation to state area i is proportional to the larger of the House amount H_i and the Senate amount S_i given by

$$H_i = \frac{35}{159} \left(\frac{P_i}{P_+} + \frac{P_i/C_i}{(P/C)_+} + \frac{U_i}{U_+} \right) + \frac{27}{159} \left(\frac{I_i}{I_+} + \frac{E_iT_i}{(ET)_+} \right)$$

(E1)

and

$$S_i = \frac{P_iE_i/C_i}{(PE/C)_+} ,$$

(E2)

where

P_i population;
U_i urbanized population;
C_i per capita income;
I_i income tax amount, which is the median of three values: $0.01L_i$, $0.15K_i$, and $0.06L_i$;
L_i federal individual income tax liabilities;
K_i state individual income tax collections;
T_i net state and local tax collections;
E_i tax effort, equal to T_i/R_i;
R_i total personal income.

These represent data elements provided by the Bureau of the Census and other agencies of the U.S. Department of Commerce; the subscript plus sign signifies summation over the subscript. For example, $P_+ = \Sigma_i P_i$, $(P/C)_+ = \Sigma_i(P_i/C_i)$, $(ET)_+ = \Sigma_i E_i T_i$, and $(PE/C)_+ = \Sigma_i P_i E_i/C_i$.

The fractions 35/159 and 27/159 result from the fact that the legislation dictates that the allocation is the amount to which the state would be entitled if one third of $3.5 billion were allocated among states on the basis of population (P_i/P_+), urbanized population (U_i/U_+), and population inversely weighted for per capita income (($P_i/C_i)/(P/C)_+$) and if one half of $1.8 billion were allocated among states on the basis of each of income tax collection (I_i/I_+) and general tax effort ($E_iT_i/(ET)_+$). When it is possible, simplified but correct statements of the formulas are presented.

State area i's portion of the total GRS funds allocated for an entitlement period can be written as X_i/X_+, where X_i is the maximum of S_i or H_i.[1] The size of the total amount allocated was essentially fixed for entitlement periods beginning before January 1, 1977. For later entitlement periods the total allocation size is determined on the basis of federal individual income tax collections to a maximum of $6.85 billion per year; the total allocations in these later entitlement periods have in fact been at the maximum.

The allocation to each state area is divided, in the ratio of approximately 1:2, between the state government and all local governments in the state.[2] The allocation to all local governments is called the "local share."

The local share is then divided among all county areas proportionally by the product of the county area's tax effort and population divided by per capita income. The tax effort of a county area is defined as the ratio of all "adjusted" (nonschool) taxes collected by the county and subcounty governments to the product of the county's population and per capita income. Observe that the population factors cancel, so that the proportion of the local share going to county area j is

$$\frac{\{(D_{ij} + D_{ij+})/(P_{ij}C_{ij})\}P_{ij}}{C_{ij}}\left(\frac{1}{G_i}\right) = \frac{D_{ij} + D_{ij+}}{(C_{ij})^2}\left(\frac{1}{G_i}\right) \qquad \text{(E3)}$$

where

D_{ij} adjusted taxes of county government j;

D_{ijk} adjusted taxes of subcounty government k in county j ($D_{j+} = \Sigma_k D_{ijk}$);

P_{ij} population of county area j;

C_{ij} per capita income of county area j;

G_i $\Sigma_j[(D_{ij} + D_{ij+})/(C_{ij})^2]$.

---

[1] For Alaska and Hawaii the procedure is somewhat different. In order to account for generally higher price levels, "noncontiguous State adjustment factors," say F_{Alaska}, F_{Hawaii}, are determined on the basis of the percentage of basic pay received by federal employees in those states as an allowance under Section 5941 of Title 5, U.S. Code. For entitlement periods 1–9 the factors were 1.25 for Alaska and 1.15 for Hawaii. For entitlement period 10 the factor for Hawaii increased to 1.175, while the factor for Alaska remained at 1.25. For i denoting Alaska and Hawaii the final state allocation is increased by the fraction $F_i - 1$. This augmentation is taken from a source of funds earmarked "noncontiguous States adjustment amounts." If the source does not contain enough money, the adjustments are scaled down.

[2] The ratio varies slightly from state to state because of the effect of maximum and minimum provisions.

Although population does not enter explicitly in (E3), the county area allocations are constrained by maximum and minimum provisions so that population does determine some substate allocations. No county area allocation, on a per capita basis, is permitted to exceed 145 percent or fall below 20 percent of two thirds of the state area allocation, on a per capita basis. The proportion of the total local share allocated to a county area so constrained is thus equal to 1.45 (or 0.20) times the proportion of the state area population residing in the county area, P_{ij}/P_i.

The partitioning of a county area allocation among local jurisdictions in the county takes place in several stages. Each Indian tribe or Alaskan native village with members residing in the county is allocated a fraction of the county area allocation equal to its proportion of the county area population. Next, the remainder of the county area allocation is partitioned proportionally among the county government, the ensemble of all township governments (if any), and the ensemble of all place and municipality governments by the respective amounts of adjusted taxes of the three types of governments. The total for township governments is then allocated among the individual townships so that the share for township k' is

$$\frac{D_{ijk'}}{(C_{ijk'})^2} \; \frac{1}{G_{ij}} \, , \tag{E4}$$

where $C_{ijk'}$ is the per capita income of township k' and G_{ij} equals the sum over all townships k of $D_{ijk}/(C_{ijk})^2$. The total for all place and municipality governments is partitioned analogously.

The following maximum and minimum provisions apply to all local governments:

1. No local government's allocation may exceed, on a yearly basis, 50 percent of the sum of its revenue from adjusted taxes and intergovernmental transfers.

2. No local government may receive, on a per capita basis, more than 145 percent or less than 20 percent of two thirds of the state area allocation, on a per capita basis.

3. Any local government allocation of less than \$200 a year shall be forfeited by the locality and given to the county government.

The population of a subcounty area—like that of a county—does not enter into the formula unless provision 2 applies. The number of subcounty jurisdictions affected by the constraints is indicated in Table E-1.

TABLE E-1   Data Used to Determine General Revenue Sharing Allocations, by Size of Place, Entitlement Period 6 (1975-1976)

	Number of Subcounty Jurisdictions			
Effective Constraint	Population to 2,499	Population 2,500- 9,999	Population 10,000+	Total
None	16,111	3,765	2,154	22,030
Population[a]	9,330	1,809	686	11,825
Adjusted taxes and intergovern- mental transfers of revenue[b]	1,301	253	172	1,726
Below $200 minimum payment	564	0	0	564
TOTAL	27,306	5,827	3,012	36,145

[a] Allocations at the 145- or 20-percent constraint.
[b] Allocations at the 50-percent constraint.

SOURCE: Calculations provided by the Data and Demography Division of the Office of Revenue Sharing.

Note that less than one third of the units' allocations were affected by population.

The sequence in which the maximum and minimum provisions are applied is important. The algorithm that calculates the allocations is complicated and iterative and is not described here. The law provides some flexibility for determining the allocations. If the Secretary of the Treasury decides that the data mentioned above will not "provide for equitable allocations," the Secretary may "use such additional data (including data based on estimates) as may be provided for in the regulations" (P.L. 92-512 Section 109(a)(7)). The law does not require that current estimates of population, income, or other parameters be produced, but only that if they are produced, they should be used.

The statute also allows the states to choose among several alternative intrastate allocation formulas (P.L. 92-512, Section 108(c)). To date, no states have chosen this option. Section 108(b)(5) authorizes the Secretary to determine the subcounty allocations to places, municipalities, and townships of populations not greater than 500 solely on the basis of the fraction of county population residing in the subcounty unit. Such very small places would thus be treated analogously to Alaskan native villages and Indian tribes. No measures of per capita income, adjusted taxes, or intergovernmental transfers of revenue would be needed for those small areas.

EFFECTS OF DATA ERRORS ON GENERAL REVENUE SHARING ALLOCATIONS

The effects of data errors on GRS allocations have generated much attention and misunderstanding. It should be noted that differential errors rather than uniform errors in data distort the allocations. Uniform errors do not distort the allocation because data elements in the GRS formulas appear not as totals but as proportions of larger area totals.[3] For example, state population figures enter only as fractions of the national population. Thus uniform relative errors in the population estimates are unimportant. Similarly, underestimating all per capita income in the nation by the same proportion causes no errors in the determination of GRS allocations.

Differential errors, however, are important. If per capita income is underestimated in one state and perfectly estimated elsewhere, then that one state's allocation will be too high and the allocations to the rest of the states will be too low (because the total allocation is fixed). Similarly, if per capita income is underestimated in one county (or subcounty unit within a county) and perfectly estimated elsewhere within the state (or county), then that one county's (or subcounty unit's) allocation will be too high and the allocations to the rest of the counties in the state (or unconstrained subcounty units in the county) will be too low.

The role of population data in the allocation process is often misunderstood. In practice, estimates of population are quite irrelevant to the allocations for most local areas. Only areas subject to the 20-percent or 145-percent constraint are directly affected by errors in estimates of their population.[4] As Table E-1 shows, population estimates enter directly into the calculation of the allocation for less than one third of the subcounty areas. For these areas, local population is the most important element in the calculation of the allocation. Slightly less than two-thirds of the subcounty jurisdictions receive funds roughly in proportion to the ratio of their net nonschool tax revenues to the *square* of their per capita income, divided by the sum of these ratios over all townships or municipalities in the county (see (E4)). Thus a given percent error in the population or the per capita income estimate for a locality is *more* significant if the locality is unconstrained (so that per capita income is important) than if the locality is at the 20-percent or 145-percent constraint (so that population is im-

---

[3] The sole exception is the application of the 50-percent constraint, which limits a substate government's allocation to no more than one-half the sum of its adjusted taxes and its net receipts of intergovernmental transfers of revenue.

[4] These areas include those whose allocations were not actually affected by the constraints but that would have been affected if there were no errors in the population estimates. The number of such areas is probably not large.

portant), because per capita income is squared, while population is not. (This point is dealt with more explicitly below.)

The hierarchical structure of the GRS formula insulates the effects of errors in data for different geographic levels. "Hierarchical" refers to the sequential determination of allocations to different geographic levels: first the total pie is divided among the state areas, then each state area's allocation is divided among the county areas, and then each county area's allocation is divided among the subcounty units. Thus errors in substate data of one state have no effect on allocations within another state. Similarly, errors in the per capita income estimates for units within one county cause no errors in the allocations within other counties.[5]

These aspects of the GRS program are most fortuitous for the Census Bureau. Since errors in substate data for one state do not affect the data or allocations in any other state, the Census Bureau is free to use different methods for substate areas in different states. The Census Bureau takes partial but not full advantage of this situation. Thus the Bureau uses different kinds of data for the ratio-correlation method estimates of county populations in different states; it also uses locally prepared estimates of county and subcounty populations in some states but not in others. There is no statistical reason for the Census Bureau to use the same methods to estimate the characteristics of counties or subcounty units in different states. (For discussion of uniformity of procedures, see section 5.2b of the report.)

For detailed understanding of how data errors affect the GRS allocations it is useful to examine formulas that explicitly relate errors in data to errors in allocation. Because the formulas typically are complicated we restrict our attention to two examples; see Spencer (1980) and Robinson and Siegel (1979) for more development. The first example illustrates the effect of error in subcounty per capita income estimates on allocations to unconstrained subcounty jurisdictions. The second example analyzes the effect of error in subcounty population estimates on allocations to subcounty jurisdictions whose allocations are determined according to population data. These jurisdictions are those at the 20-percent or 145-percent constraint.

EXAMPLE 1

The portion of the county area share allocated to a township (or municipality) *i* within the county, but not at a 145-, 20-, or 50-percent constraint,

---

[5] Here we ignore the possibility that errors in a locality's per capita income estimate might cause its allocation to become constrained (or unconstrained) while it would not be if there were no error in the per capita income estimate. The number of areas in which this occurs is not believed to be large.

is calculated to be proportional to the adjusted taxes for township (or municipality) i divided by the square of the per capita income for township (or municipality) i. The relative error in this share is approximately[6]

$$(d_i - \bar{d}) - 2(c_i - \bar{c}), \tag{E5}$$

where d_i is the relative error of the estimate of adjusted taxes, c_i is the relative error of the estimate of per capita income, and $\bar{d}$ and $\bar{c}$ are weighted averages of the relative errors:

$$\bar{d} = \Sigma w_j d_j \qquad \bar{c} = \Sigma w_j c_j$$

(Σ denotes summation over all townships (or municipalities) j in the county). For an arbitrary township (or municipality) k, the weights are defined as follows:

$$w_k = (D_k/C_k)^2/(\Sigma D_j/C_j^2),$$

where D_k and C_k denote the actual adjusted taxes and per capita income of township (or municipality) k.

Note that uniform relative errors in the subcounty estimates cancel: if x is added to each d_j or c_j, then x is also added to $\bar{d}$ or $\bar{c}$, so that the relative error in the share of the county area allocation (equation (E5)) is unaffected. Differential errors, such as $d_i - \bar{d}$ or $c_i - \bar{c}$, are what matters. (This result does not depend on the approximation leading to (E5); see (E4).)

Note that errors in subcounty estimates within one county do not affect the distribution of the county area allocation for another county because the relative errors and weights discussed above all pertain to jurisdictions within one county.

EXAMPLE 2

If a county or subcounty unit i is at the 145-percent (or 20-percent) constraint, then its proportionate share of the total substate allocation equals 1.45 (or 0.20) times the fraction its population is of the total of all subcounty areas in the state. The relative error in this share is approximately

$$K(p_i - p), \tag{E6}$$

---

[6] See Spencer (1980) for derivation.

where $K = 1.45$ (or 0.20), p_i is the relative error of the population estimate for subcounty unit i, and p is the relative error of the estimate of total population in the state. As before, note that uniform relative errors are irrelevant, for if p_i and p are both increased by x, the relative error (E6) is unchanged. Differential errors $p_i - p$ are important.

Unlike per capita income, subcounty population estimates within one county can affect the allocations to subcounty units within other counties in part because of the complicated manner in which the constraints are implemented. Roughly, the allocations to all county and subcounty areas at the 145- or 20-percent constraint are made first, then the remainder is distributed to unconstrained county areas on the basis of adjusted taxes and per capita income data, and then the allocations to unconstrained places within county areas are made. Also, the allocation to an area constrained at the 145- or 20-percent level is determined by the ratio of its population to the total population of all GRS areas in the state, so errors in the estimate for any one of the local areas in the state can affect the allocation to any other area subject to a 145- or 20-percent constraint. For example, if the population estimate for a local area subject to a 145- or 20-percent constraint is too low, then the allocation to that area will be too low, and the error in the allocation will be distributed to areas (county and subcounty) not at the 145- or 20-percent constraints.

Formulas similar to (E1) and (E2) can be derived for allocations to all levels of geography and for all the various constrained or unconstrained situations, but for reasons of space such formulas are not presented here. Such formulas are invaluable not only for insight but for detailed analysis. Approximate biases and variances of allocations can be calculated from such formulas and from estimates of the biases and variances of the various data elements. Using those approximations, one can analyze how errors in data can be expected to affect errors in allocations; for example, one can construct confidence intervals for the allocations. The alternative to a stochastic analysis (Spencer, 1980), as described above, is to use simulations to study the effects of data error on the allocations (Siegel, 1975; Siegel et al., 1977; Stanford Research Institute, 1974; Strauss and Harkins, 1974) (the latter two studies consider only the effects of population undercount). In the simulation approach, differences between allocations under alternative sets of data are studied.

In summary, errors in state-level data elements have substantial impact on substate allocations. Since substate data are used merely to divide a state's allocation, any error in the allocation to the state must be borne by all the substate units. Analysis by Siegel (1975) shows that at the state level, errors in the censal estimates of per capita income are more significant (i.e., cause more dollars to be misallocated) than those in the population estimates. Since errors in the estimates of postcensal change are worse for per capita income than population, this same relationship holds for postcensal estimates of population and income. Spencer's (1980) work suggests that per capita income errors are also more significant (i.e., cause more dollars to be misallocated) than population at the substate level. Since the allocations to substate units are more often based on per capita income than on population, errors in per capita income estimates also substantially affect more jurisdictions than do errors in population estimates.

REFERENCES

Bowditch, B., Horowitz, L., Jones, T., Pash, J., and Yates, J. (1974) *Overview of Distribution of Revenue Sharing Funds.* Rockville, Md.: Westat, Inc.

Nathan, R. P., Manvel, A. D., and Calkins, S. E. (1975) *Monitoring Revenue Sharing.* Washington, D.C.: The Brookings Institution.

Office of Revenue Sharing (1973 et seq.) *General Revenue Sharing. Final Data Elements.* Washington, D.C.: U.S. Department of the Treasury.

Robinson, J. G., and Siegel, J. S. (1979) Illustrative assessment of the impact of census underremuneration and income underreporting on revenue sharing allocations at the local level. *1979 Proceedings of the Social Statistics Section of the American Statistical Association.* Washington, D.C.: American Statistical Association.

Siegel, J. S. (1975) *Coverage of Population in the 1970 Census and Some Implications for Public Programs.* Bureau of the Census, Current Population Reports, Series P-23, No. 56. Washington, D.C.: U.S. Department of Commerce.

Siegel, J. S., Passel, J. S., Rives, N. W., Jr., and Robinson, J. G. (1977) *Developmental Estimates of the Coverage of the Population of States in the 1970 Census: Demographic Analysis.* Bureau of the Census, Current Population Reports, Series P-23, No. 65. Washington, D.C.: U.S. Department of Commerce.

Spencer, B. D. (1980) *Benefit-Cost Analysis of Data Used to Allocate Funds.* Lecture Notes in Statistics 3. New York: Springer-Verlag.

Stanford Research Institute (1974) *General Revenue Sharing Data Study.* Vol. III, Vol. IV. Menlo Park, Calif.: Stanford Research Institute.

Strauss, R. P., and Harkins, P. B. (1974) The impact of population undercounts on General Revenue Sharing allocations in New Jersey and Virginia. *National Tax Journal* XXVII:617-624.

U.S. Congress, Joint Committee on Internal Revenue Taxation (1973) *General Explanation of the State and Local Fiscal Assistance Act and the Federal-State Tax Collection Act of 1972.* Washington, D.C.: U.S. Government Printing Office.

A Note on the Use of Postcensal Population Estimates in Employment and Unemployment Measures

BRUCE D. SPENCER

Postcensal estimates of population figure prominently in official measures of employment and unemployment for subnational areas. Their role is sketched here[1]; for further details, see Bureau of Labor Statistics (1977).

It is convenient to define and distinguish between unemployment rates and unemployment ratios: the *unemployment rate* is the number of unemployed people divided by the sum of the number of employed and unemployed people; the *unemployment ratio* is the number of unemployed people divided by the population of working age; employment rates and ratios are defined analogously. Note that the sum of the employment rate and the unemployment rate for an area equals 1.

To estimate unemployment rates for states, the Bureau of Labor Statistics uses data from the Current Population Survey (cps), state unemployment insurance (ui) records, and the decennial census. For the 10 largest states (and for New York City and Los Angeles Standard Metropolitan Statistical Areas) the unemployment rates are estimated directly on the basis of cps data. For the other states, unemployment rates are estimated with information from the cps, from the decennial census, from tax reports of employers covered by the ui program, and from other sources. The bls combines these diverse kinds of data according to complicated procedures, including the so-called "handbook" or "70-step" method; see Goldstein (1978) and National Commission on Employment and Unemployment Statistics (1979) for further description and references.

---

[1] The purpose of this brief exposition is to illustrate some uses of the postcensal population estimates and not to discuss estimation of labor force parameters; for some alternative ways of estimating unemployment for local areas, see Gonzalez and Hoza (1978).

To estimate the total *number* of employed and unemployed persons for the current month in each state, the Census Bureau multiplies the respective sample ratio estimate from the CPS by a population control. This control is derived by extrapolation from the most recent July 1 postcensal population estimate for the civilian working-age population.

The Bureau of Labor Statistics estimates the number of employed and unemployed people in each labor market area (LMA) in a state by prorating the numbers for the state in proportion to the handbook estimates of employment and unemployment for the LMA. An LMA generally consists of a central city or cities and surrounding territory within commuting distance. Each LMA comprises an integral number of counties (except in New England, where it comprises an integral number of towns).

For many LMA's, estimated total employment is allocated among the counties in the LMA in proportion to the postcensal estimates of the total population of the counties. But estimated total unemployment within an LMA is allocated among constituent counties on the basis of data other than postcensal population estimates. Total employment for many incorporated places of at least 2,500 population is estimated to equal the county employment times the ratio of the postcensal population estimate for the place to the population estimate for the county. Total unemployment for the same incorporated places is estimated on the basis of data other than postcensal population estimates. For some incorporated places, mainly those that are newly incorporated or have changed their boundaries, both employment and unemployment are derived by proration of county totals on the basis of postcensal population estimates. Since the unemployment rates for many counties and places are estimated by the ratio of the estimated total unemployment to the sum of the total employment and unemployment, the postcensal population estimates for counties and for incorporated places of at least 2,500 population affect these rates.

REFERENCES

Bureau of Labor Statistics (1977) Technical Instructions for a Revised State and Area Unemployment Estimating System. Program Memorandum SP-77-35. U.S. Department of Labor.

Goldstein, H. (1978) State and Local Labor Force Statistics. Background Paper No. 1. National Commission on Employment and Unemployment Statistics, Washington, D.C.

Gonzalez, M. E., and Hoza, C. (1978) Small area estimation with application to employment and housing estimates. *Journal of the American Statistical Association* 73(361): 7–15.

National Commission on Employment and Unemployment Statistics (1979). *Counting the Labor Force.* Washington, D.C.: U.S. Government Printing Office.

Models for Error in Postcensal Population Estimates

BRUCE D. SPENCER

INTRODUCTION AND CONCLUSIONS

The methodology underlying the postcensal population estimates is complex (see Appendix A), and a useful way to analyze the errors in these estimates is to construct models incorporating the components of error. Such models provide insight into the ways in which different kinds of errors (such as those arising from estimating a migration rate or a censal population) affect the estimates both of postcensal population and of postcensal change. We focus primarily on the effects of census undercoverage on the postcensal estimates; the error structure in the incremental administrative records method (AR) estimates is also investigated.

Several approximations are required because nonlinear functions of random variables are analyzed and because simplicity in the models is desired. An important tool for analyzing the nonlinear functions of random variables is the delta method (see Bishop et al., 1975, pp. 486ff; Keyfitz, 1968, pp. 339–340; Rao, 1973, pp. 388ff). The notation $A \doteq B$ is used to mean $A = B + \epsilon$, where ϵ is a remainder term arising from the delta method or from a simplifying assumption. The analysis is heuristic, and bounds for the remainder terms are not given.

Although the present analysis is not complete—not all of the components of error are considered, and only parts of the estimation methodology are treated—some conclusions can be drawn (discussed below):

1. The effect of census undercount on the national population update decreases slightly over time.[1] The estimate of net national increase is unaffected by census undercount.

2. The effect of census undercount on ratio-correlation (RC) estimates of state or county postcensal population does not decrease over time. For an area experiencing growth the estimate of net increase afforded by the ratio-correlation method becomes progressively more affected by census undercount as the postcensal interval gets longer.

3. Net undercoverage in the census also affects subnational estimates of net migration made by either the component method II (CM II) or the administrative records method; this effect tends to remain constant over time.

4. The AR method is better used as a multiple-increment updating procedure than as a single-increment procedure. That is, to estimate 1975 population with AR, it is better to estimate separately changes over 1970–1972 and 1972–1975 and add them than to estimate the change over 1970–1975 directly.

Future work could usefully attempt to relax the simplifying assumptions employed in this analysis; develop quantitative estimates of the moments of components of error; extend the scope of this analysis, so that more methodology is analyzed; and derive bounds for the remainder terms arising from approximations.

NOTATION

Capital letters with or without subscripts denote parameter values; estimates of the parameters are distinguished by a circumflex. Generally, the subscripts i, j, and k refer to states, counties, and subcounty areas, respectively. The letter t refers to time, measured in years, with $t = 0$ corresponding to April 1, 1970.

In particular,

$P(t)[\hat{P}(t)]$ true [estimated] total population of the U.S. at time t;

$P_i(t)[\hat{P}_i(t)]$ true [estimated] population of state i at time t;

$P_{ij}(t)[\hat{P}_{ij}(t)]$ true [estimated] population of county j in state i at time t;

$P_{ijk}(t)[\hat{P}_{ijk}(t)]$ true [estimated] population of place k in county j in state i at time t.

---

[1] By "effect" we mean contribution to the relative error in the parameter (here, total national population) being estimated.

Errors in estimates are denoted by lowercase letters:

$$p(t) = \hat{P}(t) - P(t)$$
$$p_i(t) = \hat{P}_i(t) - P_i(t)$$
$$p_{ij}(t) = \hat{P}_{ij}(t) - P_{ij}(t)$$
$$p_{ijk}(t) = \hat{P}_{ijk}(t) - P_{ijk}(t).$$

Net undercoverage rates for the censal population estimates are denoted by

$$A = -p(0)/P(0)$$
$$A_i = -p_i(0)/P_i(0)$$
$$A_{ij} = -p_{ij}(0)/P_{ij}(0)$$
$$A_{ijk} = -p_{ijk}(0)/P_{ijk}(0).$$

Jacob Siegel and colleagues at the Census Bureau have developed estimates of A and A_i, but estimates with comparable reliability of A_{ij} and A_{ijk} are not available (see Bureau of the Census, 1977).

Errors in the estimates of net increase are

$$\Delta_P(t) = p(t) - p(0) = \hat{P}(t) - \hat{P}(0) - (P(t) - P(0))$$
$$\Delta_P(i; t) = p_i(t) - p_i(0) = \hat{P}_i(t) - \hat{P}_i(0) - (P_i(t) - P_i(0))$$

and $\Delta_P(i, j; t)$ and $\Delta_P(i, j, k; t)$ similarly defined.

A BASIC DECOMPOSITION

The term "relative error" is applied to the ratio of the error to the appropriate parameter value; for example, $p(t)/P(t)$ is the relative error in the estimate of total U.S. population. It is often convenient to work with relative errors because the algebra is simple and because relative errors for different estimates may be comparable even though the parameters under estimation are not.

The error in the postcensal population estimate decomposes into the error in the estimate of net increase and the error in the censal estimate:

$$p(t) = \Delta_P(t) + p(0). \tag{G1}$$

Dividing by $P(t)$ yields

$$\frac{p(t)}{P(t)} = \frac{\Delta_P(t)}{P(t)} - A\frac{P(0)}{P(t)}. \tag{G2}$$

Clearly, the relative error in the postcensal population estimate is the sum of the updating error divided by the population size plus the net undercoverage rate times the ratio of census population to postcensal population. Similar relations hold for $P_i(t)$, $P_{ij}(t)$, and $P_{ijk}(t)$.

EFFECT OF UNDERCOUNT ON NATIONAL UPDATES

Relation (G2) suggests that if $P(t)$ increases over time and $\Delta_P(t)$ is unaffected by census undercoverage, then as the postcensal interval gets longer, the effect of undercount decreases as the population increases (in percentage terms) since the last census. In fact, $\hat{P}(t)$ increases over time, and $\Delta_P(t)$ is not affected by undercoverage in the previous census because the estimate of net national increase is essentially derived from reported data on births, deaths, and net immigration since the previous census. We conclude that the effect of census undercount on the national postcensal population estimate decreases over time as the population increases (in percentage terms).

EFFECT OF UNDERCOUNT ON SUBNATIONAL UPDATES

Analogues to (G1) and (G2) hold for the subnational errors $p_i(t)$, $p_{ij}(t)$, and $p_{ijk}(t)$. However, at the subnational level one cannot conclude that the relative errors in the postcensal estimates become progressively less affected by undercoverage in the previous census. The explanation is to be found only partly in the declines in population experienced by some subnational areas (including Rhode Island, New York, Pennsylvania, and the District of Columbia; see Bureau of the Census (1979)). The more interesting fact is that for some methods, differential undercoverage in the census affects the subnational estimates of net increase.

RATIO-CORRELATION METHOD (RC)

For the postcensal estimates obtained by the ratio-correlation method (RC), the effect of undercoverage does not change over time. Consider the RC estimate of postcensal population for county j in state i. Defining the actual and estimated shares X_{ij} and $\hat{X}_{ij}$ by

$$X_{ij} = \frac{P_{ij}(t)/P_{ij}(0)}{P_i(t)/P_i(0)} \qquad \hat{X}_{ij} = \frac{\hat{P}_{ij}(t)/\hat{P}_{ij}(0)}{\hat{P}_i(t)/\hat{P}_i(0)}, \tag{G3}$$

where $\hat{X}_{ij}$ is obtained with the use of a regression equation (see Appendix A), $P_{ij}(t)$ is estimated by

$$\hat{P}_{ij}(t) = \hat{X}_{ij}\hat{P}_{ij}(0)\hat{P}_i(t)/\hat{P}_i(0). \tag{G4}$$

Application of the delta method to (G4) yields

$$\frac{p_{ij}(t)}{P_{ij}(t)} \doteq \frac{x_{ij}}{X_{ij}} + \frac{p_{ij}(0)}{P_{ij}(0)} + \frac{p_i(t)}{P_i(t)} - \frac{p_i(0)}{P_i(0)} \tag{G5}$$

where lowercase letters denote errors, e.g., $x_{ij} = \hat{X}_{ij} - X_{ij}$. Relation (G5) can also be expressed as

$$\frac{p_{ij}(t)}{P_{ij}(t)} \doteq \frac{x_{ij}}{X_{ij}} + \frac{p_i(t)}{P_i(t)} - A_{ij} + A_i. \tag{G6}$$

Comparing (G6) with (G2), notice that in (G6), unlike (G2), the coefficients of the undercoverage terms A_i and A_{ij} do not change over time.

To see the effect of census undercoverage on the RC estimates of net increase, one can subtract $p_{ij}(0)/P_{ij}(t)$ from both sides of (G5) and rearrange terms to obtain

$$\frac{\Delta_P(i, j; t)}{P_{ij}(t)} \doteq \frac{x_{ij}}{X_{ij}} + \frac{\Delta_P(i; t)}{P_i(t)} - \frac{P_{ij}(t) - P_{ij}(0)}{P_{ij}(0)}A_{ij} + \frac{P_i(t) - P_i(0)}{P_i(0)}A_i. \tag{G7}$$

If the proportional growth for the county equals that for the state, say,

$$\lambda = \frac{P_{ij}(t) - P_{ij}(0)}{P_{ij}(t)} = \frac{P_i(t) - P_i(0)}{P_i(t)},$$

then (G7) becomes

$$\frac{\Delta_P(ij; t)}{P_{ij}(t)} \doteq \frac{x_{ij}}{X_{ij}} + \frac{\Delta_P(i; t)}{P_i(t)} + \lambda(A_i - A_{ij}). \tag{G8}$$

Relations (G7) and (G8) show the effect of undercoverage in the census on the RC estimates of net increase. In fact, if population growth is significant (because the annual rate of growth is high or because the time interval t is long), then λ can increase, and undercoverage in the census can have a progressively greater effect on the estimates of net increase. Similar relationships hold for state estimates.

COMPONENT METHOD II AND ADMINISTRATIVE RECORD METHOD ESTIMATES OF NET INMIGRATION

This section considers estimation of net migration for a county by either the component method II (CM II) or the administrative records (AR) method. Because these methods treat the population over age 65 separately, it is convenient to use the term "elderly" to refer to any person born at least 65 years before the estimate date; any person who is not elderly is called "young." Let the actual and estimated net migration rates over the postcensal period be denoted by R_{ij} and $\hat{R}_{ij}$ for county j in state i. The actual and estimated net numbers of migrants to state i (to county j in state i) over the postcensal period are represented by M_i and $\hat{M}_i$ (M_{ij} and $\hat{M}_{ij}$). Also, let Y_i and $\hat{Y}_i$ (Y_{ij} and $\hat{Y}_{ij}$) denote the actual and estimated young populations in state i (county j in state i) on the census date. (The difference $Y - \hat{Y}$ equals the undercount.)

It is convenient to assume that R_{ij}, $\hat{R}_{ij}$, M_i, and $\hat{M}_i$ are all nonzero, that there is no international migration, and that there are no group quarters or military populations. We also assume, for illustrative purposes, that actual and estimated net natural increases are zero.

The estimates $\hat{R}_{ij}$ are obtained from the matching of tax returns (in AR) or from changing patterns in school enrollments (in CM II). Because the migration estimates for counties are adjusted to state totals in a complicated way (see Appendix A, sections 3.4e and 3.8), the estimates $\hat{M}_{ij}$ do not generally equal $\hat{Y}_{ij}\hat{R}_{ij}$. The "unadjusted" estimates of postcensal population, $\hat{Y}_{ij}(1 + \hat{R}_{ij})$, are scaled by a factor $\hat{\gamma}_i$ to equal the state estimate $\hat{Y}_i + \hat{M}_i$:

$$\hat{\gamma}_i = \frac{\hat{Y}_i + \hat{M}_i}{\sum_j \hat{Y}_{ij}(1 + \hat{R}_{ij})}.$$

The estimate of net migration for county l, say, is estimated as a residual:

$$\hat{M}_{il} = \hat{Y}_{il}(1 + \hat{R}_{il})\hat{\gamma}_i - \hat{Y}_{il}$$

$$= \frac{\hat{Y}_{il}(1 + \hat{R}_{il})(\hat{Y}_i + \hat{M}_i)}{\sum_j \hat{Y}_{ij}(1 + \hat{R}_{ij})} - \hat{Y}_{il}.$$

The actual net number of migrants to county l is given by

$$M_{il} = Y_{il}R_{il}$$

$$= \frac{Y_{il}(1 + R_{il})(Y_i + M_i)}{\sum_j Y_{ij}(1 + R_{ij})} - Y_{il}.$$

Suppose, for simplicity, that the actual and estimated net migration rates R_{ij} and $\hat{R}_{ij}$ are constant over counties, so that

$$\hat{M}_{il} = \frac{\hat{Y}_{il}(\hat{Y}_i + \hat{M}_i)}{\sum_j \hat{Y}_{ij}} - \hat{Y}_{il}$$

$$= \frac{\hat{Y}_{il}\hat{M}_i}{\hat{Y}_i}$$

and, similarly, that

$$M_{il} = Y_{il}(M_i/Y_i).$$

Using the delta method, we can easily obtain

$$\frac{\hat{M}_{il} - M_{il}}{M_{il}} \doteq \frac{m_i}{M_i} + \left(\frac{y_{il}}{Y_{il}} - \frac{y_i}{Y_i}\right), \tag{G9}$$

where lowercase letters represent errors, e.g., $m_i = \hat{M}_i - M_i$. Clearly, the relative error in the county estimate of net migration arises partly from error in the state estimate of net migration and partly from the differential (across counties) census undercoverage of the young population. Moreover, the two components of error are equally important.

If we now allow the net migration rates R_{ij} and $\hat{R}_{ij}$ to vary over counties, we can in like manner derive

$$\frac{\hat{M}_{il} - M_{il}}{Y_{il} + M_{il}} \doteq F_i \cdot \frac{m_i}{M_i} + \left(F_{il} \cdot \frac{y_{il}}{Y_{il}} - F_i \cdot \frac{y_i}{Y_i}\right) - \left(F_{il} \cdot \frac{r_{il}}{R_{il}} - F_i \cdot \frac{r_i}{R_i}\right) \tag{G10}$$

where the weights F_i and F_{il} are defined by $F_i = M_i/(Y_i + M_i)$ and $F_{il} = M_{il}/(Y_{il} + M_{il})$ and where the state migration rate and error are $R_i = M_i/Y_i$ and $r_i = \hat{R}_i - R_i$. Here the error in the county estimate arises from error in the state estimate of net migration, from a (weighted) under-coverage differential for the young populations, and from a (weighted) differential between the relative error for the county migration rate and that for the statewide average. If the county and the state grow at the same rate, the effect of census undercoverage of the estimate of net migration for the county remains constant over time.

Net natural increase affects the estimates of net inmigration in two ways: the adjustment factor $\hat{\gamma}_i$ has a more complicated form, and the "base" population by which the migration rate $\hat{R}_{ij}$ is multiplied is no longer $\hat{Y}_{ij}$ but rather $\hat{Y}_{ij}$ plus one-half the net natural increase of the young population. The presence of group quarters populations or international migration would also affect the estimates of net inmigration (for the first reason above). However, the methods used to derive (G9) and (G10) can still be applied to decompose the error in the net inmigration estimate when such other components of change are taken into account. The decompositions are straightforward but tedious to derive and are not given here.

SINGLE-INCREMENT VERSUS MULTIPLE-INCREMENT ADMINISTRATIVE RECORDS ESTIMATES

The administrative records method is generally used as a multiple-increment updating procedure. For example, the Census Bureau obtained estimates of 1975 population by separately estimating population changes from 1970 to 1972 and 1972 to 1975 and adding the sum of these changes to the 1970 censal population estimate. A single-increment method would have estimated the change from 1970 to 1975 in one step. The question of whether a multiple-increment updating procedure is superior to a single-increment updating procedure is examined analytically below. The analysis suggests that the multiple-increment updating procedures are superior (have smaller bias and variance) to the single-increment procedures. The analysis is heuristic, however, and utilizes many simplifying assumptions. Future research could examine the sensitivity of the conclusion to modifications of the assumptions. In particular, empirical studies comparing updates by the two kinds of procedures should be done.

Let times 0, s, t satisfy $0 < s < t$, where 0 is the time of the last census, t is the time of the current postcensal estimate, and s is the time of an intermediate postcensal estimate. The multiple-increment procedure estimates population change over $[0, t)$ by separately estimating changes over $[0, s)$ and $[s, t)$, while the single-increment procedure estimates change over $[0, t)$ directly. For simplicity of presentation we ignore births, deaths, international immigration, and special populations. Thus all population change arises from net internal migration. Consider

P actual population at time 0;
S actual number of nonmovers (stayers);
I actual number of inmigrants;
E actual number of outmigrants;
M actual net number of inmigrants ($M = I - E$).

Notice that $P = S + E$. Denote the counts of nonmovers, inmigrants, and outmigrants provided by the matching of IRS tax returns by S', I', and E' and denote the estimate of P by $\hat{P}$. Now let the random variables C, c_I, and c_E be defined by

$$S'/S = C$$
$$I'/I = C(1 + c_I)$$
$$E'/E = C(1 + c_E).$$

Here C is the coverage ratio for nonmovers and c_I and c_E are the relative deviations of the coverage ratios of inmigrants and outmigrants from C. Let $p = \hat{P} - P$ be the error in $\hat{P}$.

Now consider the AR estimate $\hat{M}$ of M,

$$\hat{M} = \frac{I' - E'}{S' + E'}\hat{P}, \tag{G11}$$

and note that

$$\hat{M} = \frac{IC(1 + c_I) - EC(1 + c_E)}{SC + EC(1 + c_E)} \cdot P(1 + p/P)$$

$$= \frac{M + Ic_I - Ec_E}{P(1 + Ec_E/P)} \cdot P(1 + p/P).$$

Application of the delta method gives

$$\frac{\hat{M} - M}{M} \doteq c_I I/M - c_E(1 + M/P)E/M + p/P. \tag{G12}$$

For simplicity of analysis we assume that c_I, c_E, and p are mutually uncorrelated. The relative variance of $\hat{M}$ is thus given by

$$\text{Var}\left(\frac{\hat{M}}{M}\right) \doteq \left(\frac{I}{M}\right)^2 \sigma_I^2 + \left(\frac{E}{M}\right)^2 (1 + M/P)^2\sigma_E^2 + \sigma_P^2, \quad \text{(G13)}$$

where σ_I^2, σ_E^2, and σ_P^2 are the variances of c_I, c_E, and p, respectively. The relative variance of the postcensal population estimates $\hat{P} + \hat{M}$ is given by

$$\text{Var}\left(\frac{\hat{P} + \hat{M}}{P + M}\right) \doteq \left(\frac{I}{P + M}\right)^2 \sigma_I^2 + \left(\frac{E}{P + M}\right)^2 (1 + M/P)^2\sigma_E^2 + \sigma_P^2.$$

$$\text{(G14)}$$

We can now compare the variances of the single-increment and double-increment procedures. Let subscripts $0s$, $0t$, and st refer to the ends of the intervals $[0, s)$, $[0, t)$, and $[s, t)$; for example, M_{st} denotes the actual net number of migrants over the period $[s, t)$. Similarly, σ_{Ist}^2 and σ_{Est}^2 are the variances of c_I and c_E for matches of tax returns between times s and t. The relative variances of the estimates of population at times 0 and s are written σ_{P0}^2 and σ_{Ps}^2. We assume that c_{Ist} and c_{Est} are not correlated with $\hat{P}_s$.

It follows from (G14) that the relative variance of the single-increment update is given by

$$\text{Var}\left(\frac{\hat{P}_{0t} + \hat{M}_{0t}}{P_t}\right) \doteq \left(\frac{I_{0t}}{P_t}\right)^2 \sigma_{I0t}^2 + \left(\frac{E_{0t}}{P_t}\right)^2 (1 + M_{0t}/P_0)^2\sigma_{E0t}^2 + \sigma_{P0}^2.$$

$$\text{(G15)}$$

To derive the variance of the double-increment update, we first note that the relative variance of $\hat{P}_s$ is

$$\sigma_{Ps}^2 \doteq \left(\frac{I_{0s}}{P_s}\right)^2 \sigma_{I0s}^2 + \left(\frac{E_{0s}}{P_s}\right)^2 (1 + M_{0s}/P_0)^2\sigma_{E0s}^2 + \sigma_{P0}^2. \quad \text{(G16)}$$

As in (G13) the relative variance of $\hat{M}_{st}$ is

$$\text{Var}\left(\frac{\hat{M}_{st}}{M_{st}}\right) \doteq \left(\frac{I_{st}}{M_{st}}\right)^2 \sigma_{Ist}^2 + \left(\frac{E_{st}}{M_{st}}\right)^2 (1 + M_{st}/P_s)^2\sigma_{Est}^2 + \sigma_{Ps}^2,$$

$$\text{(G17)}$$

and the relative variance of the double-increment update is thus

$$\text{Var}\left(\frac{\hat{P}_s + \hat{M}_{st}}{P_t}\right) \doteq \left(\frac{I_{st}}{P_t}\right)^2 \sigma_{Ist}^2 + \left(\frac{E_{st}}{P_t}\right)^2 (1 + M_{st}/P_s)^2 \sigma_{Est}^2 + \sigma_{Ps}^2.$$

$$\text{(G18)}$$

The variance of the single-increment update minus that of the double-increment update is approximately

$$(I_{0t})^2 \sigma_{I0t}^2 + (E_{0t})^2(1 + M_{0t}/P_0)^2 \sigma_{E0t}^2 - \left(\frac{I_{0s}P_t}{P_s}\right)^2 \sigma_{I0s}^2 - (I_{st})^2 \sigma_{Ist}^2$$

$$- \left(\frac{E_{0s}P_t}{P_s}\right)^2 (1 + M_{0s}/P_0)^2 \sigma_{E0s}^2 - (E_{st})^2(1 + M_{st}/P_s)^2 \sigma_{Est}^2.$$

$$\text{(G19)}$$

For simplicity, suppose there is no outmigration, that inmigration is linear over time, and that P is large in comparison to I; i.e., $0 = E_{0s} = E_{0t} = E_{st}$, $I_{0t} = tI$, $I_{0s} = sI$, $I_{st} = (t - s)I$, and P_t/P_s is approximately unity. Then (G19) is approximately equal to

$$I^2\{t^2\sigma_{I0t}^2 - s^2\sigma^2_{I0s} - (t - s)^2\sigma_{Ist}^2\}. \qquad \text{(G20)}$$

For many specifications of σ_{I0t}^2, σ_{I0s}^2, and σ_{Ist}^2, expression (G20) will be positive, indicating that the double-increment procedure has (subject to the assumptions above) smaller variance. For example, if σ_{Ist}^2 has the form $a + b(t - s)^d$ for positive constants a and d and nonnegative b, then (G20) is positive.

The biases in the two procedures can be analyzed in similar fashion. The analog of (G20) for biases is

$$I(t\mu_{I0t} - s\mu_{I0s} - (t - s)\mu_{Ist}), \qquad \text{(G21)}$$

where μ_I is the mean of c_I. If μ_{Ist} is constant over possible values of s and t, then (G21) is zero, so the biases in the single- and multiple-increment updates are approximately the same. If μ_{Ist} has the form $a + b(t - s)^d$, where b and d are positive (negative) constants, then the biases in both updating procedures will be positive (negative), but the absolute bias in the single-increment update will be larger. Generally, the absolute bias of the single-increment update is believed to be at least as large as that of the double-increment update.

The analysis above suggests that the multiple-increment procedure is superior to the single-increment procedure. Future research should relax some of the simplifying assumptions used, especially that of the lack of covariances. It would be especially useful to compare the conclusions of this analysis with the results of empirical tests of accuracy of the two kinds of updates.

REFERENCES

Bishop, E., Fienberg, S., and Holland, P. (1975) *Discrete Multivariate Analysis.* Cambridge, Mass.: MIT Press.

Bureau of the Census (1977) *Developmental Estimates of the Coverage of the Population of States in the 1970 Census.* Current Population Reports, Series P-23, No. 65. Washington, D.C.: U.S. Department of Commerce.

Bureau of the Census (1979) *Revised 1977 and Provisional 1978 Estimates of the Population of States and Components of Change.* Current Population Reports, Series P-25, No. 799. Washington, D.C.: U.S. Department of Commerce.

Keyfitz, N. (1968) *Introduction to the Mathematics of Population.* Reading, Mass.: Addison-Wesley.

Rao, C. R. (1973) *Linear Statistical Inference and Its Applications.* 2nd ed. New York: John Wiley and Sons.

Evaluation Design and the Use of Low-Precision Benchmarks

CARL N. MORRIS

If one is fortunate enough to know the "true values," say, $\theta_1, \ldots, \theta_n$, of some parameters in n areas, then one can use those values to evaluate alternative estimators of them. If one of the estimators takes the values $\mathbf{Y} = (Y_1, \ldots, Y_n)$, then a measure of the accuracy of $\mathbf{Y}$ is

$$L(\theta, \mathbf{Y}) = \sum_1^n (Y_i - \theta_i)^2.$$

Other loss functions are possible, of course, as was noted in section 3.1 of the report. For example, it may be appropriate to weight the squares in the above formula to reflect the size of an area, etc., but these points are ignored here in order to illustrate the issues of interest.

Usually, θ is unobservable, but instead one has $\theta_1^*, \ldots, \theta_n^*$, which are other estimates of $\theta_1, \ldots, \theta_n$ and have their own variances. Assume that, independently,

$$\theta_i^* \sim N(\theta_i, \nu_i), \qquad i = 1, \ldots, n.$$

If $\nu_i = 0$, then $\theta_i^* = \theta_i$, but otherwise, θ_i^* is unbiased for θ_i with variance $\nu_i > 0$. If ν_i is small in comparison to the variance of Y_i, one might call θ_i^* a "high-precision" estimate or benchmark, while larger values of ν_i would have "low-precision" (high variance).

So long as θ_i^* is independent of $\mathbf{Y}$, then given $\mathbf{Y}$,

$$EL(\theta^*, \mathbf{Y}) = \sum_1^n E(Y_i - \theta_i^*)^2 \,|\, Y_i = \sum_1^n [(Y_i - \theta_i)^2 + \nu_i]$$

so that one cannot estimate the true loss for **Y** by substituting θ_i^* (or θ_i) in $L(\theta, \mathbf{Y})$. Instead one might use the unbiased estimator

$$L^*(\theta^*, \mathbf{Y}) = \sum_1^n \{(Y_i - \theta_i^*)^2 - \nu_i\}$$

of $L(\theta, \mathbf{Y})$.

How else must we account for the ν_i? Two strategies often are followed for areas with large ν_i. Strategy 1 ignores all areas in the evaluation with ν_i exceeding some threshold; strategy 2 uses the areas with large ν_i by clustering them so that the combination of areas has sufficiently small variance. I wish to observe that neither of these strategies is optimal, although each may be convenient and appropriate at times.

Strategy 1 violates the principle of sufficiency. Additional information can always be used profitably, even if the information is quite imperfect. In this case, ignoring areas with low-precision benchmarks is inferior to using them, but with lower weight. The appropriate weights depend on the variability of the Y_i as well as the V_i.

For example, suppose $\nu_i = V_1$ for $i = 1, \ldots, N_1$ and $\nu_i = V_2 > V_1$ for $i = N_1 + 1, \ldots, N_1 + N_2$. Then the first N_1 areas are (relatively) high precision, the last N_2 are low precision, and $N_1 + N_2 = n$ areas that are available for evaluation. Suppose $Y_i \sim N(\theta_i, W)$ with W unknown. Then W measures the precision of Y_i, and if $\theta_1, \ldots, \theta_n$ were known ($V_1 = V_2 = 0$), one would compute

$$L(\theta, \mathbf{Y}) = \sum_1^n (Y_i - \theta_i)^2$$

which estimates nW. When the θ_i^* must be used, it is better to use the mixed estimator

$$\frac{\alpha}{N_1} \sum_1^{N_1} [(Y_i - \theta_i^*)^2 - V_1] + \frac{1 - \alpha}{N_2} \sum_{N_1+1}^n [(Y_i - \theta_i^*)^2 - V_2]$$

to estimate W than to use $(1/N_1)\Sigma[(Y_i - \theta_i^*)^2 - V_1]$ alone, provided one chooses α to be the optimal value:

$$\alpha = Q_2/(Q_1 + Q_2),$$

where

$$Q_1 = \text{Var} \frac{1}{N_1} \sum_1^{N_1} [(Y_i - \theta_i^*)^2 - V_1] = \frac{2}{N_1} (W + V_1)^2$$

and

$$Q_2 = \text{Var}\, \frac{1}{N_2} \sum_{N_1+1}^{N_2} [(Y_i - \theta_i{}^*)^2 - Y_2] = \frac{2}{N_2}(W + V_2)^2.$$

(W must be known to compute this, but the maximum likelihood estimate of W can be used in this manner and causes no undue complication.) Thus the low-precision estimates are as useful as the high-precision estimates, provided

$$N_2 = \left(\frac{W + V_2}{W + V_1}\right)^2 N_1,$$

and the relative efficiency of a low-precision estimate in this example is

$$\text{Effic} = \left(\frac{W + V_1}{W + V_2}\right)^2.$$

Note that if V_2 is small with respect to W, low-precision estimates are nearly as good as high-precision ones. We usually expect V_1 to be considerably less than W and V_2 to be on the order of W. Then low-precision estimates would have about one-fourth the efficiency of high-precision estimates, but the value of such information cannot be denied. More general examples can be constructed. The main point is that strategy 1, which ignores low-precision estimates, can be costly, especially if such estimates outnumber high-precision estimates.

The second evaluation strategy, strategy 2, pools several low-precision estimates to produce fewer high-precision ones. This method is efficient if areas having the same mean are grouped. Otherwise, serious biases within these groupings will go undetected because only the *average* performance of the various Y_i associated with the groupings will be evaluated. Low-precision estimates may frequently correspond to small areas, so willingness to use low-precision benchmarks may provide genuine benefits for evaluation of small-area estimates.

The comments made in this section fall under the general heading of developing good evaluation strategies. The Panel has recommended that the Bureau of the Census evaluate its procedures whenever possible. The Census Bureau should consider carefully its methods for making evaluations, for this is not a well-charted terrain, and the utility of an evaluation cannot exceed the quality of the evaluation strategy.

Effect of Biases in Census Estimates on Evaluation of Postcensal Estimates

BRUCE D. SPENCER

Error in the decennial and special census estimates of population and income confounds the evaluation of both estimates of postcensal level and of postcensal change. This error arises from net census undercoverage and from underreporting of income. This appendix focuses on the effects of undercoverage bias on the population estimates, although analysis of biases of income estimates would be similar. In particular, this appendix delineates precisely how undercount affects the evaluations based on decennial or special censuses:

1. Use of the difference between the postcensal estimate and the (special) census count to estimate the error in the former underestimates this error because it ignores the bias in the special census counts.

2. Use of the difference between the postcensal estimate and the (special) census count to estimate the error in the estimate of postcensal change is also affected by undercoverage, but to a lesser degree. Even if the undercoverage rates for the base-year census (at the beginning of the postcensal period) and the (special) census used for evaluation are the same, the undercoverage will affect the evaluations.

Consider estimation for an arbitrary geographic unit and for time t, with

P_t true value of population;

$\hat{P}_t$ postcensal estimate of population;

$\tilde{P}_t$ census estimate (decennial or special) of population.

232

Time $t = 0$ refers to the previous decennial census, and by convention, $\hat{P}_0$ equals $\tilde{P}_0$. The undercoverage rate α_t is defined by

$$\alpha_t = \frac{P_t - \tilde{P}_t}{P_t},$$

and the error in the postcensal estimate of net change, Δ_t, is given by

$$\Delta_t = \hat{P}_t - \hat{P}_0 - (P_t - P_0).$$

Estimates of statewide net undercoverage rates for 1970 range from less than zero (estimated net overcount for Wisconsin and Utah) to more than 0.07 (for New Mexico, Arkansas, and Alaska) (see Bureau of the Census, 1977, Table VII-D). Of course, substate rates cannot vary less than state rates.

The usual evaluation studies of postcensal estimates use the difference between the postcensal estimate and a current census estimate,

$$\hat{P}_t - \tilde{P}_t, \tag{I1}$$

to measure error in either the estimate of postcensal level, $\hat{P}_t$, or the estimate of postcensal change, $\hat{P}_t - \hat{P}_0$. In general, (I1) does not provide a good estimate of the error in the estimate of level,

$$\hat{P}_t - P_t, \tag{I2}$$

because of undercoverage in $\tilde{P}_t$. To see this, note that (I1) can also be expressed as

$$\hat{P}_t - P_t + \alpha_t P_t,$$

which shows that the net undercoverage $\alpha_t P_t$ may well exceed the actual error, (I2).

For example, in section 2.2 of the report, the Panel analyzes the accuracy of county estimates by using the average over counties of the absolute relative differences

$$|\hat{P}_t - \tilde{P}_t|/\tilde{P}_t \tag{I3}$$

between postcensal estimates and special census estimates to support inferences about the expected value of

$$|\hat{P}_t - P_t|/P_t. \tag{I4}$$

The average of (I3) for 133 counties with special censuses taken during 1974–1976 ranged from 0.039 to 0.064 for different methods (see Table 2.3). However, one must be guarded in inferring that the mean value of (I4) lies in this range, because the magnitude of α_t is often comparable to or larger than (I3). In other words, unless the undercoverage rate for an area is much smaller than the relative difference, (I3), between the estimate and the census count, the value of (I3) will tend to grossly overestimate the true relative absolute error (I4).

To avoid this problem, one might consider using (I1) to estimate Δ_t or using (I3) to estimate $|\Delta_t|/P_t$. In general, this use of (I1) or (I3) is less sensitive to the presence of undercoverage than is the use of (I1) for estimating (I2) or the use of (I3) for estimating (I4). Even if one could assume that undercoverage is constant over time—that α_0 and α_t are the same—however, the use of (I1) and (I3) for making inferences about Δ_t would still in fact be sensitive to the level of undercoverage. Better inferences about the properties of Δ_t can be made by taking undercoverage into consideration.

Some decompositions of error will be useful. Observe that

$$\hat{P}_t - \tilde{P}_t = \Delta_t + \alpha_t P_t - \alpha_0 P_0. \tag{I5}$$

Letting

$$\epsilon_t = P_t(\alpha_t - \alpha_0) \qquad \gamma_t = \alpha_0(P_t - P_0),$$

one obtains

$$\hat{P}_t - \tilde{P}_t = \Delta_t + \gamma_t + \epsilon_t. \tag{I6}$$

Here ϵ_t is the effect of differences between the base year decennial census undercoverage rates and the later special (or decennial) census undercoverage rates, and γ_t is the joint effect of population change and decennial census undercoverage in the base year. Little is known about how undercoverage rates in special censuses compare to those in decennial censuses. Studies (Bureau of the Census, 1973) indicate that the national undercoverage rates for the 1960 and 1970 decennial censuses differed by 0.002, but how much the rates for subnational areas changed is unknown. On the other hand, we do know that γ_t can be large for places that are growing or declining substantially and that have moderate-to-large undercoverage rates.

Let us now assume $\epsilon_t = 0$ and focus on γ_t. A possible way to improve the estimates of Δ_t is to remove the estimated effect of γ_t. Notice that

$$\tilde{P}_t - \tilde{P}_0 = (1 - \alpha_0)(P_t - P_0) - \epsilon_t,$$

so that if an estimate $\hat{\alpha}_0$ of α_0 is available, one can estimate γ_t by $\hat{\gamma}_t$, where

$$\hat{\gamma}_t = \frac{\hat{\alpha}_0(\tilde{P}_t - \tilde{P}_0)}{1 - \hat{\alpha}_0}.$$

More sophisticated estimates of γ_t could also be developed. Thus the estimated effect of γ_t would be removed if instead of using $\hat{P}_t - \tilde{P}_t$ to estimate Δ_t one used

$$\hat{P}_t - \tilde{P}_t - \hat{\gamma}_t. \tag{17}$$

A major difficulty with $\hat{\gamma}_t$ is the inaccuracy of the estimates $\hat{\alpha}_0$. At the state level these estimates are questionable, and at substate levels they are worse. Nevertheless, for most places one can be fairly confident that α_0 is positive, say, $\alpha_0 \geq 0.01$. For these places a cautious estimate of γ_t would be

$$\frac{0.01(\tilde{P}_t - \tilde{P}_0)}{0.99}.$$

This is preferable to the present implicit use of $\hat{\gamma}_t = 0$. Alternative approaches could estimate α_0 for a substate area by an estimate for the state or for the nation as a whole. Since the population-weighted average of undercoverage rates for substate areas equals the state undercoverage rate, a simple but reasonable approach consists of estimating α_0 for a substate area by the estimate for the whole state. In fact, these estimates of γ_t could be substantial and have a significant effect on evaluations of methods. This effect is believed to be greatest when an evaluation compares the postcensal estimate against a decennial census count, that is, when 10-year updates are evaluated.

For example, consider evaluating the postcensal estimates for Florida counties. The estimates of proportional growth over 1970–1976 were more than 0.24 for more than half of Florida's 77 counties. Extrapolating, one could suppose that for t referring to April 1, 1980, $\tilde{P}_t - \tilde{P}_0 > 0.4\tilde{P}_0$ for many counties. For Florida, $\hat{\alpha}_0 \geq 0.05$, so that $\hat{\gamma}_t/\tilde{P}_t$ could be near 0.016. This value is large: the difference between the usual measures of accuracy (the average of (I3) for all counties) for alternative estimation procedures may well be less than 0.016.

A possible result of correcting for $\hat{\gamma}_t$ is a more realistic sensitivity to bias. For example, if for fast growing places an estimating methodology is unbiased and has small variance, use of (I1) rather than (I7) to study the errors could lead to inferences that the estimates were biased upward.[1] In this case, (I1) would be primarily estimating γ_t rather than Δ_t. For another example, suppose that for a given class of areas, two methods had biases with opposite signs—method A tended to underestimate and method B to overestimate. In this example, use of (I1) instead of (I7) will make method A appear better than it really is and will make method B appear worse than it really is.

In making inferences about the relative errors Δ_t/P_t one may similarly divide (I7) by $\tilde{P}_t/(1 - \hat{\alpha}_0)$ rather than by $\tilde{P}_t$. However, this extra adjustment for undercoverage will generally have less impact than adjustment for $\hat{\gamma}_t$. In particular, division by $1/(1 - \hat{\alpha}_0)$ will have negligible impact on comparisons between methods when $\hat{\alpha}_0$ is constant for all places.

Some empirical study is needed to compare the measures (I1) and (I7) over a range of $\hat{\gamma}_t$ values. How sensitive to $\hat{\gamma}_t$ are the estimates of accuracy for different methodologies? Do the rankings of the methodologies change? Of course, the accuracy of $\hat{\gamma}_t$ adjustments rests on that of the estimates of undercoverage, but by adjusting for undercoverage (as discussed above) one can expect some improvement in evaluation, given current knowledge about undercoverage.

This analysis has focused on population estimates; income estimates can be handled similarly by drawing on knowledge of the effects of income underreporting and population undercoverage on income estimates. For both population and income, understanding how biases in census estimates affect the evaluation aids interpretations of the evaluations.

REFERENCES

Bureau of the Census (1973) *Estimates of Coverage of Population by Sex, Race, and Age: Demographic Analysis.* Evaluation and Research Program PHC(E)-4. Washington, D.C.: U.S. Department of Commerce.

Bureau of the Census (1977) *Developmental Estimates of the Coverage of the Population of States in the 1970 Census: Demographic Analysis.* Current Population Reports, Series P-23, No. 65. Washington, D.C.: U.S. Department of Commerce.

[1] In fact (see section 2.2), it appears that estimates for fast growing places are biased downward. Because these conclusions are based on the use of (I1) rather than (I7), it is possible that the estimates for fast growing places have even more downward bias than evaluations indicate.

Stabilization by Empirical Bayes Methods

CARL N. MORRIS

The evaluation methods considered in sections 3.2 and 5.2 of the report are used to choose different procedures or to determine how to average two (or more) procedures. We have recommended that procedures be used that "best" predict sample data, for example, from independent Current Population Surveys. Sometimes good weights can be determined from data without an independent data set for evaluation. Fay and Herriot (1979) have identified such an application for the Census Bureau and showed the method works well for estimating per capita income in small areas in a census year. The method uses empirical Bayes modeling approaches to generalize Stein's estimator appropriately for that application (see Efron and Morris, 1975). We discuss the Fay-Herriot application here and suggest other census uses of this methodology. We then consider the relationship between empirical Bayes weights and weights determined from regression methods.

In a census year, income is measured imperfectly for all areas because it is determined from a sample. Assuming that good sampling procedures have been followed, we consider the sample mean of income in each area as an unbiased estimate of the mean per capita income for the area. Let Y_i be the sample mean in the ith area. In small areas, even accounting for finite population corrections, variances will be quite large. (A census would alleviate the problem of large variances, of course, but in 1970 20-percent samples were used, and the 1980 census will collect data on income for not more than 50 percent of the population in small areas.) Let V_i be the variance of the sample mean Y_i.

Instead of using the sample mean directly the Census Bureau can regress the sample income estimates of small areas on other characteristics (given by the matrix $\times$) correlated with income (Fay and Herriot take these to be IRS income and housing values) to derive an income predictor for each area, for example,

$$\hat{Y}_i = \mathbf{X}_i' \hat{\beta} \tag{J1}$$

with $\mathbf{X}_i'$ the ith row of $\times$. This "regression predictor," being estimated from many degrees of freedom, has small variance. But unbiasedness cannot be guaranteed, as Fay and Herriot showed for the 1970 census.

In decennial census years, both "sample estimators" and "regression predictors" of the preceding paragraphs are available for all areas. Such is not the case in postcensal years and possibly will not be the case in 1985. How then, in a census year, should one choose between the unbiased, but noisy, "sample estimators," Y_i and the (probably) biased, but low-variance, "regression predictors," $\hat{Y}_i$?

An empirical Bayes estimator estimates the true mean in the ith area, $\mu_i = EY_i$, by

$$\hat{\mu}_i = (1 - B_i)Y_i + B_i\hat{Y}_i, \tag{J2}$$

with $B_i = V_i/(V_i + W)$, W being the variance of μ_i about the regression surface. W itself may be estimated from the data (see Efron and Morris, 1975, 1977; Fay and Herriot, 1979). If W is large, B_i is close to zero, so that $\hat{\mu}_i$ is nearly Y_i, the sample mean. Small values of W put almost full weight on $\hat{Y}_i$, the regression estimate. Formula (J2) is a Bayes estimator, but since W is estimated and not determined independently of the data, it is called empirical Bayes (see Efron and Morris, 1975, 1977; Fay and Herriot, 1979; and below).

Statisticians have long known how best to average independent unbiased estimators: they weight each by its reciprocal variance. With one of the two estimators, Y_i, being unbiased, statisticians have developed the theory, under an assumed model of the error distribution of $\hat{Y}_i$, for estimating the mean squared error of $\hat{Y}_i$. The resultant estimator weights both independent estimators Y_i and $\hat{Y}_i$ by the reciprocals of their mean squared errors. Carried out formally, this procedure results in an empirical Bayes estimator. It reduces to Stein's celebrated estimator provided, for example, that the sample mean of income is equally variable in every area, which would not be true in Census Bureau applications.

The Panel endorses the work of Fay and Herriot and encourages continued use of such methodology for income estimation. We believe the method will work in other applications when small-area estimates must be made. The Panel does not seek widespread use of such methods—they do not apply in most situations and will not necessarily be beneficial in all applicable cases[1]—but empirical Bayes methods are likely to improve estimators used in a variety of cases. Some possible applications are presented briefly below; these ideas are suggested for future research.

In making the estimates of 1973 subcounty population used for revenue sharing, the Census Bureau set migration rates for areas with less than 1,000 people equal to the county migration rate because the Bureau had little faith in the AR estimates of small-area migration rates. An empirical Bayes estimator could have been used to produce weighted averages of small-area rates and county rates and almost surely would have been superior to the county rates used.

Empirical Bayes methods should also be explored as alternatives to the "tolerance check" methods currently used for estimating subcounty migration rates (see Appendix A, section 4.1d). In the tolerance check approach, if the coverage ratio for tax returns of a subcounty area differs by more than a given amount from the coverage ratio for the county, the subcounty migration rate is estimated by either the county migration rate or the migration rate for a group of subcounty areas. Empirical Bayes methods could be used to determine weights for averaging the initial subcounty migration rates and the migration rates for the group of subcounties. These improved migration estimates could then be used in the AR estimates.

The errors in local area population estimates vary by characteristics or "covariates" of the area, such as population size, growth rate, region of the county, etc. To control for these covariates when evaluating the accuracy of the estimates, a common technique uses two-way or higher-dimensional cross-classifications of average error by strata of values of population size, percent change, etc. (see Tables 2.1–2.11). When the number of observations in a cell is small, this "stratification" analysis may be unreliable. An alternative approach is to fit linear or other models to express the error in the estimate as a linear (or other) function of the covariates (percent change, population size, etc.). These covariance

---

[1] For example, as shown below, the resulting estimates of population for the small areas will generally be biased. For areas with net migration rate below (above) the county rate, the bias will be positive (negative). In some cases the magnitude of the bias can be large (see Rao and Schinozaki, 1978). If the population estimates are used to determine the allocations of funds to the areas for successive time periods, areas with small net migration rates (relative to the county) get a favorable treatment in the long run at the expense of areas with larger net migration rates; and the situation is even more severe when the migration rates for the small areas are set equal to the county rate. An evaluation is recommended before adoption of empirical Bayes methods in any particular application, to be sure that improvements will occur.

models can aid in understanding how errors tend to vary according to the characteristics of an area. However, this practice carries some risk if the model specification fits poorly. Empirical Bayes methods could be used to shrink the stratification estimates toward estimates produced by the covariance model, thereby stabilizing the stratification estimates and reducing the risk of the covariance model approach.

Two different methods of empirically determining weights for averaging different estimators have been suggested:

1. Regression estimators (see section 5.2 of the report) can be used when two or more estimators have been proposed if CPS or special census data are available for a current year. The observations used for the evaluation must be approximately unbiased, but they need be available only in a small portion of all areas. This method works best for determining postcensal estimates.

2. Empirical Bayes estimates can be used to combine unbiased sample estimates with regression predictors for the areas. This method can only be applied in sampled areas, unlike method 1, but it does not require CPS or special census data. Its prime purpose is to improve estimates made in a census year, as in the Fay-Herriot application. The two methods apply to different situations. The Panel has recommended that method 2 also be used to stabilize and improve postcensal estimates, but empirical Bayes methods do not perform the evaluation function of method 1.

REFERENCES

Efron, B., and Morris, C. (1975) Data analysts using Stein's estimator and its generalizations. *Journal of the American Statistical Association* 70(350):311–319.

Efron, B., and Morris, C. (1977) Stein's paradox in mathematical statistics. *Scientific American* (May):119–127.

Fay, R., and Herriot, R. (1979) Estimates of income for small places: An application of James-Stein procedures to census data. *Journal of the American Statistical Association* 74(366, Part 1):261–277.

Rao, C. R., and Shinozaki, N. (1978) Precision of individual estimates in simultaneous estimation of parameters. *Biometrika* 65(1):23–30.

Recommendation for Question on Residence: Letter to the Director of the Office of Revenue Sharing

The Panel's report includes a recommendation that a place of residence question be included in the 1980 IRS individual income tax returns. This recommendation repeats one made by the Panel in January 1979 with reference to the 1979 IRS individual income tax returns. The Panel's letter to the director of the Office of Revenue Sharing, reproduced below, explains the importance of the recommendation; technical details about the place of residence question can be found in Appendix A, section 4.1d.

NATIONAL RESEARCH COUNCIL
ASSEMBLY OF BEHAVIORAL AND SOCIAL SCIENCES

2101 Constitution Avenue   Washington, D. C. 20418

COMMITTEE ON NATIONAL STATISTICS

January 9, 1979

Dr. Bernadine Denning
Director
Office of Revenue Sharing
Department of the Treasury
2401 E Street, N.W.
Columbia Plaza High Rise
Washington, D.C. 20226

Dear Dr. Denning:

The Panel on Small-Area Estimates of Population and Income has recently been established at the request of the Bureau of the Census under the auspices of the National Academy of Sciences. The Panel is in the process of reviewing the procedures used by the Bureau of the Census to make postcensal estimates of population and income for small areas. These estimates are used for the allocation of general revenue sharing funds, as well as for other major public purposes, such as health planning. Although the study will not be completed until December 1979, the Panel is writing to urge that the 1979 IRS income tax returns contain a special question to determine exact place of residence, as was included on 1975 tax returns, for use by the Bureau of the Census.

The information reported on the tax returns plays an essential role in the estimation procedure. By comparing changes in address and income of specific individuals in two sets of tax records, the Bureau uses the information on the tax returns in its estimation of migration and changes in per capita income. The mailing address on the return often is insufficient for determining in which unit of local government the filer of the return actually resides. A question on residence was asked on the 1975 IRS returns. It provided the essential information for allocation of mailing addresses to the appropriate places of residence and has served as the basis for such allocations since then. But localities experience different rates of growth, and, in many instances, the use of the 1975 allocation factors is no longer appropriate. Annexations and boundary changes are frequent and, for many places, the allocations based on city boundaries as of 1975 are no longer valid.

There is another important reason for including the place of residence question on the 1979 returns. There is a question as to how much the migration patterns and rates of change in income differ between

The National Research Council is the principal operating agency of the National Academy of Sciences and the National Academy of Engineering to serve government and other organizations

Dr. Bernadine Denning
January 9, 1979
Page 2

the populations covered and not covered by tax returns. The proportions of population covered by tax returns (i.e. either filing or claimed as an exemption on a return) vary widely from one place to another. In using the IRS data to estimate migration and changes in per capita income, the Bureau assumes that the migration patterns and the rates of change in wage and salary income are identical for the populations covered and not covered by tax returns. If the accuracy of the small-area estimates is to be significantly improved, these operational assumptions need to be evaluated and modified accordingly.

Because the filing dates for the 1979 tax returns are so close to the decennial census date of April 1, 1980, a rare opportunity exists to examine the assumptions by using the 1980 census results to compare the characteristics of the populations covered and not covered by tax returns. The Panel notes that if the residence question is deferred to another year, the ability of the Bureau to examine its assumptions will be restricted. Under the provisions of Title 13, U.S.C., confidential treatment of the data is assured.

> The Panel recommends that the place of residence question be included on the 1979 tax returns and that funds be sought to enable the Bureau of the Census to process the data obtained from the question.

The Panel is fully aware of the efforts to keep the tax form simple and to minimize the amount of non-tax information called for. We also realize that processing the responses to the question is an expensive operation. But obtaining and analyzing the responses to the question is the most practical way to get the needed information. The 95-percent response rate in 1975 indicates good public cooperation in answering such a question. If the responses to the question are not obtained and analyzed, the Bureau's ability to maintain the accuracy of the local estimates for the 1980's will be impaired and desired improvements will be impeded.

A similar letter is being sent to the Director of the Bureau of the Census. We will also send copies of the letter to the Secretary of Commerce and to the Commissioner of the Internal Revenue Service.

We would welcome the opportunity for further discussion.

Sincerely yours,

Evelyn Kitagawa

Evelyn M. Kitagawa
Chairman
Panel on Small-Area Estimates of
Population and Income

Biographical Sketches of Panel Members and Staff

EVELYN M. KITAGAWA is professor of sociology and director, Population Research Center, University of Chicago. Her research interests include social demography, mortality, and the methods of demography, and she has written widely in all three areas. She is a fellow of the American Statistical Association and a former president of the Population Association of America and of the Sociological Research Association. She obtained a B.A. in mathematics from the University of California at Berkeley and a Ph.D. in sociology from the University of Chicago.

PAUL DEMENY is vice president of The Population Council and director of the Council's Center for Policy Studies. Previously, he served on the faculty of Princeton University and was a professor of economics at the University of Michigan and at the University of Hawaii, where he was also director of the East-West Population Institute. His research interests include the economic aspects of population growth and population policy, as well as methods of demography. He is charman of the editorial committee of *Population and Development Review*. He obtained a B.A. from the University of Budapest, studied at the Institut Universitaire de Hautes Etudes Internationales in Geneva, Switzerland, and received a Ph.D. in economics from Princeton University.

EUGENE P. ERICKSEN is sampling statistician and study director at the Institute for Survey Research and assistant professor of sociology at Temple

University. Previously, he taught courses in statistics and in sociology at the Balham and Tooting College of Commerce in London, England, and at the University of Michigan. His research interests include the methodology of postcensal population estimates for local political units and the study of rural and urban communities and ethnic groups. He is a member of the American Statistical Association, the American Sociological Association, and the Population Association of America. He obtained a B.S. in mathematics from the University of Chicago and an M.A. in mathematical statistics and a Ph.D. in sociology from the University of Michigan.

CARL N. MORRIS is professor of statistics, department of mathematics, University of Texas. He previously was a senior statistician at The Rand Corporation and faculty member of Rand Graduate Institute for Policy Analysis, teaching statistics and data analysis. He has taught statistics at a number of other universities, has engaged in statistical consulting work, and has written widely on statistical methods, especially on methods of estimation and data analysis. He is a member of the American Statistical Association and the Institute of Mathematical Statistics and has served on the editorial boards of both associations. He received a B.S. in aeronautical and mechanical engineering from the California Institute of Technology and an M.S. and Ph.D. in statistics from Stanford University.

RICHARD F. MUTH is professor of economics at Stanford University and was formerly at Washington University (St. Louis), Chicago University's Graduate School of Business, and Vanderbilt University. He has been a consultant for the Institute for Defense Analysis, a member of the Presidential Task Force on Urban Renewal, 1969, and a visiting senior fellow to The Urban Institute. His research interests include urban and regional economics and price theory, especially applied to housing, and the spatial pattern of economic activities in cities. He is a member of the American Economic Association, the American Statistical Association, the Econometric Society, and the Regional Science Association. He received an A.B. and M.A. from Washington University (St. Louis) and a Ph.D. in economics from the University of Chicago.

DONALD E. PURSELL is director, Bureau of Business Research, and professor of business administration at the University of Nebraska. Formerly, he was director, Center for Manpower studies; professor of management,

Memphis State University; project specialist, The Ford Foundation; and senior research fellow, Nigerian Institute of Social and Economic Research, University of Ibadam. His research interests include labor mobility, earnings, and the relationship of population growth and economic development. He is a member of the American Economic Association, the Southern Economics Association, and the Population Association of America. He received a B.A. and M.A. from Southern Illinois University and a Ph.D. in economics from Duke University.

C. R. RAO is university professor of statistics at the University of Pittsburgh. He spent most of his professional life at the Indian Statistical Institute engaged in research and training, and he became its director in 1972. He has written widely on statistical methods and statistical inference, with particular application to biometric research. He has been president of the International Biometric Society, the International Statistical Institute, the Institute of Mathematical Statistics, and the Indian Econometric Society. He is a fellow of the Royal Society and the American Statistical Association, among other professional organizations, and an honorary fellow of the Royal Statistical Society and the American Academy of Arts and Sciences. He received an M.A. in mathematics from the University of Andhra, an M.A. in statistics from the University of Calcutta, and a Ph.D. and a Sc.D. from Cambridge University.

HARRY M. ROSENBERG is chief of the Mortality Statistics Branch, U.S. National Center for Health Statistics. Formerly, he was senior research associate at the Carolina Population Center, University of North Carolina, while also serving as adjunct professor of biostatistics and lecturer in sociology. Earlier he directed the state Office of Economic Opportunity, Ohio Department of Economic and Community Development, and served as a research fellow at the Battelle Memorial Institute in Columbus. He is interested in demographic methods and their application to planning, administration, and public policy and in the organization of statistical systems. He is a fellow of the American Statistical Association and a member of the American Public Health Association, the Population Association of America, and the Southern Regional Demographic Group. He received an A.B. in anthropology from the University of North Carolina and a Ph.D. in sociology from Ohio State University.

CONRAD TAEUBER is associate director of the Center for Population Research of the Kennedy Institute of Ethics at Georgetown University in

Washington, D.C. He was formerly associate director of the Bureau of the Census with responsibility for demographic surveys and censuses. He received a B.A., an M.A., and a Ph.D. from the University of Minnesota, the latter in 1931. Prior to coming to Washington he taught at Mt. Holyoke College. He served in the Works Progress Administration and in the U.S. Department of Agriculture and was chief statistician of the Food and Agriculture Organization of the United Nations. His work in relation to statistical surveys and censuses has included assignments with the United Nations and the Inter-American Statistical Institute. He is chairman of the Committee on National Statistics.

T. JAMES TRUSSELL is associate professor of economics and public affairs and faculty associate, Office of Population Research, Princeton University. His principal research interests are demographic methods, fertility, and family planning, and he has published research papers in all three areas. He is a member of the Population Association of America and the International Union for the Scientific Study of Population. He received a B.S. in mathematics from Davidson College, a B.Phil. from Oxford University in economics, and a Ph.D. in economics from Princeton University.

BRUCE D. SPENCER, who served as study director for the Panel's work, is now assistant professor of education statistics and policy, School of Education, Northwestern University. His major research interests include the application of statistical theory and methods to problems of public data collection and organization, uses of statistics for public policy, and demographic estimation. He is a member of the American Statistical Association, the Institute of Mathematical Statistics, the Royal Statistical Society, and the Population Association of America. He obtained a B.S. from Cornell University, an M.S. from Florida State University, and a Ph.D. in statistics from Yale University.

OF